EVALUATING EFFECTIVENESS OF PRIMARY PREVENTION OF CANCER

INTERNATIONAL AGENCY FOR RESEARCH ON CANCER

The International Agency for Research on Cancer (IARC) was established in 1965 by the World Health Assembly, as an independently financed organization within the framework of the World Health Organization. The headquarters of the Agency are at Lyon, France.

The Agency conducts a programme of research concentrating particularly on the epidemiology of cancer and the study of potential carcinogens in the human environment. Its field studies are supplemented by biological and chemical research carried out in the Agency's laboratories in Lyon and, through collaborative research agreements, in national research institutions in many countries. The Agency also conducts a programme for the education and training of personnel for cancer research.

The publications of the Agency are intended to contribute to the dissemination of authoritative information on different aspects of cancer research. A complete list is printed at the back of this book.

INTERNATIONAL UNION AGAINST CANCER

The International Union Against Cancer (UICC) is a non-governmental organization devoted exclusively to all aspects of the worldwide fight against cancer. Founded in 1933, it now has 254 member organizations in 84 countries, including cancer leagues and societies, cancer research and/or treatment centres and, in certain countries, ministries of health. It carries out a wide range of programmes in collaboration with hundreds of volunteer experts in various cancer-related fields. The UICC is non-profit-making, non-political, and non-sectarian; its headquarters are in Geneva, Switzerland.

NORDIC CANCER UNION

The Nordic Cancer Union (NCU), founded in 1953, is the cooperative organ for the Danish Cancer Society, the Cancer Society of Finland, the Icelandic Cancer Society, the Norwegian Cancer Society and the Swedish Cancer Society. NCU functions through meetings, annual scientific symposia and establishing Nordic collaboration in cancer control.

INTERNATIONAL UNION AGAINST CANCER

INTERNATIONAL AGENCY FOR RESEARCH ON CANCER

NORDIC CANCER UNION

EVALUATING EFFECTIVENESS OF PRIMARY PREVENTION OF CANCER

Edited by

M. Hakama, V. Beral, J.W. Cullen and D.M. Parkin

IARC Scientific Publications No. 103

International Agency for Research on Cancer
Lyon, 1990

Published by the International Agency for Research on Cancer,
150 cours Albert Thomas, 69372 Lyon Cedex 08, France

© International Agency for Research on Cancer, 1990

Distributed by Oxford University Press, Walton Street, Oxford OX2 6DP, UK

Distributed in the USA by Oxford University Press, New York

All rights reserved. No part of this publication may be reproduced, stored in a retrieval system, or transmitted, in any form or by any means, electronic, mechanical, photocopying, recording, or otherwise, without the prior permission of the copyright holder.

ISBN 92 832 2103 6

ISSN 0300-5085

Printed in the United Kingdom

CONTENTS

Foreword vii
List of participants ix
Introduction xi

Phases in cancer control: intervention research
    J.W. Cullen 1

Quantification of the effects of preventive measures
    J. Kaldor and D.P. Byar 13

Effectiveness of primary prevention of occupational exposures on cancer risk
    A.J. Swerdlow 23

Effects of changes in diet and tobacco use on risk of cancer

    Changes in tobacco consumption and lung cancer risk: evidence from national statistics
        A.D. Lopez 57

    Changes in tobacco consumption and lung cancer risk: evidence from studies of individuals
        D.R. Shopland 77

    Changes in diet and changes in cancer risk: observational studies
        D.M. Parkin and M.P. Coleman 93

Interventions on diet and tobacco use

    Prevention of cancer: review of the evidence from intervention trials
        M. Coleman and M. Law 113

    WHO European Collaborative Trial in the multifactorial prevention of coronary heart disease
        World Health Organization European Collaborative Group 123

Changes in cancer incidence in North Karelia, an area with a comprehensive preventive cardiovascular programme
T. Hakulinen, E. Pukkala, M. Kenward, L. Teppo, P. Puska, J. Tuomilehto and K. Kuulasmaa 133

A primary prevention study of oral cancer among Indian villagers. Eight-year follow-up results
P. C. Gupta, F.S. Mehta, J.J. Pindborg, D.K. Daftary, M.B. Aghi, R.R. Bhonsle and P.R. Murti 149

Primary prevention of cancer: relevant Multiple Risk Factor Intervention Trial results
W.T. Friedewald, L.H. Kuller and J.K. Ockene 157

Other selected interventions

V. Beral 171

The future

Community-based cardiovascular disease prevention programmes: models for cancer prevention?
R.A. Carleton and T.M. Lasater 179

Current issues in cancer chemoprevention
J.E. Buring and C.H. Hennekens 185

Summary 195

Index 203

FOREWORD

The prevention of cancer by reducing exposure to identified carcinogenic agents is the obvious sequel to research into the causes of the disease. However, surprisingly little effort has been applied to establishing how effective such preventive measures really are in reducing cancer incidence.

To bring our knowledge up-to-date on this important aspect of cancer control, UICC and IARC collaborated in organizing a workshop in Reykjavik, Iceland, at which methods, past results and experience from related areas of preventive intervention were reviewed and discussed. The working papers presented at this workshop have been revised and developed to form the present volume. As a product of UICC's project on Evaluation of Primary Prevention Programmes, the publication extends the UICC interest in prevention that has already resulted in numerous publications on screening, and it is also a natural extension of the major research effort at IARC over the last 20 years to identify cancer causes. This project belongs to the UICC Epidemiology and Prevention Programme, of which the chairman is Professor K. Aoki, and the workshop was organized by the members of the Project Committee, who are also the editors of this volume.

The Reykjavik workshop was sponsored by the Nordic Cancer Union, which took financial responsibility for the arrangements at the Icelandic Cancer Society, with local organization coordinated by Dr Hrafn Tulinius.

K. Aoki
Chairman, UICC Epidemiology
and Prevention Programme

L. Tomatis, M.D.
Director, IARC

LIST OF PARTICIPANTS

Dr E. Arnesen
ISM
P.B. 417
Tromsø
Norway

Dr V. Beral
ICRF Cancer Epidemiology and Clinical Trials Unit
Radcliffe Infirmary
Oxford
UK

Dr J.E. Buring
Brigham and Women's Hospital
Harvard Medical School
55 Pond Avenue
Brookline, MA 02146
USA

Dr D.P. Byar
Division of Cancer Prevention and Control
National Cancer Institute
Bethesda, MD 20892-4200
USA

Dr R.A. Carleton
Pawtucket Heart Health Program
Memorial Hospital
126 Prospect Street
Pawtucket, RI 02865
USA

Dr J. Cheney
Editorial and Publication Services
International Agency for Research on Cancer
150 Cours Albert Thomas
Lyon
France

Dr M. Coleman
Unit of Descriptive Epidemiology
International Agency for Research on Cancer
150 Cours Albert Thomas
Lyon
France

Dr J.W. Cullen
Division of Cancer Prevention and Control
National Cancer Institute
Bethesda, MD 20892-3100
USA

Dr L. Döbrössy
Cancer Unit
World Health Organization
Regional Office for Europe
8 Scherfigsvej
Copenhagen Ø
Denmark

Dr W.T. Friedewald
Division of Cancer Prevention and Control
National Cancer Institute
Bethesda, MD 20892
USA

Dr P.C. Gupta
Tata Institute of Fundamental Research
Homi Bhabha Road
Bombay – 400 005
India

Dr M. Hakama
University of Tampere
Department of Public Health
Box 607
Tampere
Finland

Dr T. Hakulinen
Finnish Cancer Registry
Liisankatu 21 B
Helsinki
Finland

Dr J. Kaldor
Unit of Biostatistics Research and Informatics
International Agency for Research on Cancer
150 Cours Albert Thomas
Lyon
France

Dr G. Lamm
Klinikum der Universität Heidelberg
Abteilung Klinische Sozialmedizin
Bergheimer Strasse 58
Heidelberg
Federal Republic of Germany

Dr M. Law
The Medical College of St Bartholomew's Hospital
University of London
Charterhouse Square
London
UK

Dr A.D. Lopez
Global Epidemiological Surveillance and Health Situation Assessment
World Health Organization
Geneva
Switzerland

Dr D.M. Parkin
Unit of Descriptive Epidemiology
International Agency for Research on Cancer
150 Cours Albert Thomas
Lyon
France

Dr R. Sasaki
Department of Preventive Medicine
School of Medicine
Nagoya University
Tsurumai-cho, Showa-ku
Nagoya 466
Japan

Dr A.J. Swerdlow
London School of Hygiene and Tropical Medicine
Keppel Street
London
UK

Dr H. Tulinius
Icelandic Cancer Society
P.O. Box 5420
125 Reykjavik
Iceland

INTRODUCTION

Research into the causation of human cancer has, as its ultimate goal, the prevention of disease. The causes of cancer may have been defined in terms of discrete chemical or biological agents or as more general aspects of human behaviour or lifestyle. In all cases, however, the notion that something is a cause of cancer implies that its change will lead to change in the corresponding effect. One might expect, therefore, that identification of cancer risk factors would be followed by studies on how to reduce the exposure, and if this is achieved, to confirm that the risk of cancer declines as expected from the cause–effect model. In fact, there have been relatively few studies of the effectiveness of different strategies in reducing cancer-causing exposures, even in terms of the exposure levels achieved by various programmes and in terms of the actual reduction in resulting cancers. At present, the most extensive evidence that a reduction in risk follows reduced exposure comes from unplanned observations, such as those of individuals or groups changing their lifestyles (particularly giving up smoking) or environment, or of changes in industrial practices.

More systematic efforts at evaluation have been made in relation to the major risk factors for cardiovascular disease. Here, several large trials based on health education, in a broad sense, have been carried out. Since the main factors examined (smoking, diet) are believed also to have important roles in cancer development, and as cancer mortality in many cases has been recorded among other causes of death, it has been possible to analyse the results of these trials to see if a reduction in cancer occurred alongside the desired reduction of cardiovascular disease.

This monograph examines from various angles the problems of evaluating primary prevention efforts. Its focus is on studies where there has been a planned reduction in a believed carcinogenic agent or behaviour, and this has been followed by measurement of cancer risk. Thus it does not consider the ascertainment of, for example, smoking reduction following anti-tobacco campaigns. Primary prevention is taken here to include all actions aimed at reducing the occurrence of cancer, but does not include activities aimed at the early detection of cancer or precursor lesions using screening methods. The effectiveness of screening programmes has been evaluated in several previous publications from the UICC Project on the Evaluation of Screening Programmes for Cancer (see below).

In this volume, some methodological aspects are first presented, in terms of the general approach to prevention research and the quantitative aspects that must be considered.

Next, various categories of preventive activities are examined. The occupational field provides good examples of reduction in exposure to specific carcinogens (though the reasons for the reduction were not always the carcinogenicity) and in some cases clear reductions in the related cancers have been documented. Next, the evidence that changes in tobacco and dietary habits can reduce cancer risk is reviewed. A first group of papers brings together observational results, showing that changes in cancer incidence do occur when diet or tobacco use is modified, even if the modification is not the result of an organized preventive action. A second group of papers examines the results of actual intervention programmes, mostly undertaken with the aim of reducing cardiovascular disease incidence, but also including an interesting example from India of an attempt to reduce oral cancers due to tobacco use. A last part of this section of the monograph describes aspects of cancer prevention in diverse fields such as chemotherapy, nuclear medicine, environmental pollution and radiation protection.

The monograph concludes with two contributions that suggest directions in which cancer preventive trials seem likely to move in the future, looking at how community-based programmes can be carried out and the possibilities for chemoprevention, prevention based on the administration of cancer-inhibiting substances, instead of on the elimination of cancer-causing agents.

A summary of the results of the studies described and of the prospects for the future is presented at the end of the volume.

Publications of the UICC Project on the Evaluation of Screening Programmes for Cancer

Screening in Cancer (1978) Miller, A.B., ed., Geneva, International Union Against Cancer

Screening for Cancer (1984) Prorok, P.C. & Miller, A.B., eds, Geneva, International Union Against Cancer

Screening for Cancer of the Uterine Cervix (1986) Hakama, M., Miller, A.B. & Day, N.E., eds, Lyon, International Agency for Research on Cancer

Screening for Breast Cancer (1988) Day, N.E. & Miller, A.B., eds, Bern, Toronto, Hans Huber

Screening for Gastrointestinal Cancer (1988) Chamberlain, J. & Miller, A.B., eds, Bern, Toronto, Hans Huber

Evaluating Effectiveness of Primary Prevention of Cancer
Ed. M. Hakama, V. Beral, J.W. Cullen & D.M. Parkin
Lyon, International Agency for Research on Cancer
© IARC, 1990

PHASES IN CANCER CONTROL: INTERVENTION RESEARCH

J.W. Cullen

*Division of Cancer Prevention and Control,
National Cancer Institute,
National Institutes of Health,
Bethesda, MD, USA*

A resurgence of interest in cancer prevention and control[1] has occurred in the United States (USA) since the passage of the National Cancer Act in 1971 (92nd Congress, 1971). This interest is derived in part from the recognition of a need for a national strategy to translate the accumulated progress in basic knowledge about cancer into nationwide benefits. In attempting to plan and implement cancer control technologies, however, a number of barriers are encountered, not the least of which is the choice of a methodological framework from which to proceed. While information related to the benefits of primary prevention (particularly tobacco-use reduction), early detection (viz., for breast and cervical cancers) and state-of-the art cancer management is available, the methodologies to implement cancer control are either lacking or deficient. In an effort to address these circumstances in the USA, the National Cancer Institute (NCI) formulated a national plan and a methodological strategy in the early 1980s to carry out broad intervention research to reduce cancer mortality rates in the USA by 50 % before the end of the century (Greenwald & Sondik, 1986).

The new plan proposed the means to test, in human populations, cancer control strategies developed from basic research findings. The plan involved five research phases: hypothesis generation, methods development, controlled intervention trials, defined population studies, and demonstration and implementation studies. This process begins with a systematic review of existing data from etiologic research (experimental and epidemiological) and randomized trials (where cancer control treatment goals are of interest), determination of available and applicable intervention methods, implementation of intervention efficacy trials, and finally the launching of large target population studies intended to have a nationwide impact on public health. All of NCI's cancer control research studies are classified into these cancer control phases. Whenever a new study is planned, the study rationale and proposed methodology are evaluated on the basis of which phase of intervention research is appropriate.

---

[1]The term 'cancer control' will be used throughout this paper as a substitute for 'cancer prevention and control'.

History of cancer control

Breslow and his colleagues (1977) have noted that the early leadership for organized cancer control activities in the USA was provided by surgeons who routinely encountered cancer. In 1913, a time when cancer of the uterine cervix was fatal in the USA, despite the appreciation that early detection and prompt treatment could yield a cure, a group of surgeons from the American Gynecological Society formed two organizations that were to play crucial roles in cancer control. They were the American Society for the Control of Cancer, which became the American Cancer Society (ACS) in 1945, and the American College of Surgeons which first developed standards for cancer clinics in 1930.

Partly due to the action by some states to establish facilities and services for the care and treatment of people with cancer, Congress acted in 1937 to form the NCI. Among the charges, Congress (75th Congress, 1937) mandated that the NCI promote 'the useful application of ... [research]... results with a view to the development and prompt widespread use of the most effective methods of prevention, diagnosis, and treatment of cancer'. In the following three decades, however, the cancer control objectives set out in the Act were difficult to achieve. This difficulty was largely due to the limited scientific base applicable to cancer control and the lack of a systematic approach to identify, test and implement effective technologies.

An organizational milestone occurred when Congress again emphasized cancer control with the passage of the National Cancer Act of 1971. For the first time, a national collaborative effort was mounted through the private and public sectors. Cancer control became a distinct programme concerned with identifying, testing, evaluating and promoting the application of all means to combat cancer. However, the new cancer control programme was cast as a demonstration programme, implying that effective intervention methodologies already existed and were ready for broad-scale implementation when, in fact, they often were not available or were still under development.

The new strategy for cancer control research

The NCI began a comprehensive effort in 1982 to review and reorganize priorities and programmes in cancer control (Greenwald & Cullen, 1984, 1985) and outlined the scientific basis of a national programme. The process explained the assumptions underlying basic laboratory and clinical research. Further, cancer control was defined as 'the reduction of cancer incidence, morbidity, and mortality through an orderly sequence, from research on interventions and their impact in defined populations to the broad, systematic application of the research results'. Given the confusion historically associated with the content and intent of cancer control as a priority in a national programme, an explanation of the principal elements of the definition follows:

(a) The *goal* of cancer control is to reduce cancer rates: incidence, morbidity and mortality rates.

(b) The phrase, *'through an orderly sequence'* means using the scientific method systematically from basic or clinical investigations to a broad application of the research results in large target populations.

(c) Cancer control must involve *interventions*. An epidemiological study that examines an etiologic factor but does not set into motion an intervention to affect public health would, therefore, not be considered cancer control.

(d) The phrase *'impact in defined populations'* is included because research on cause-and-effect relationships does not necessarily reveal how to achieve an effective, wide-scale impact in large populations or in samples of populations from which one can extrapolate to large populations.

(e) Finally, to have a public health impact, cancer control efforts are achieved through the *broad, systematic application of the research results* in large target populations.

Phases of cancer control research

To guarantee objectivity in the conduct of cancer control, to approach cancer control research systematically, and to facilitate the development of interventions after basic research findings have been collected, a decision-making model comprising an orderly sequence of five phases of research was proposed (Figure 1). Phase I of this model develops hypotheses; Phase II identifies and/or develops the methodologies to test the hypotheses in comparison or controlled trials; Phases III and IV involve implementing efficacy trials that test the stated hypotheses using the identified methodological approaches; and Phase V applies the results of the efficacy studies as demonstration studies in large target populations. Between each phase and the next, there is a 'decision point' with criteria to determine if the results from one phase justify proceeding to the next.

Figure 1. Phases of cancer control

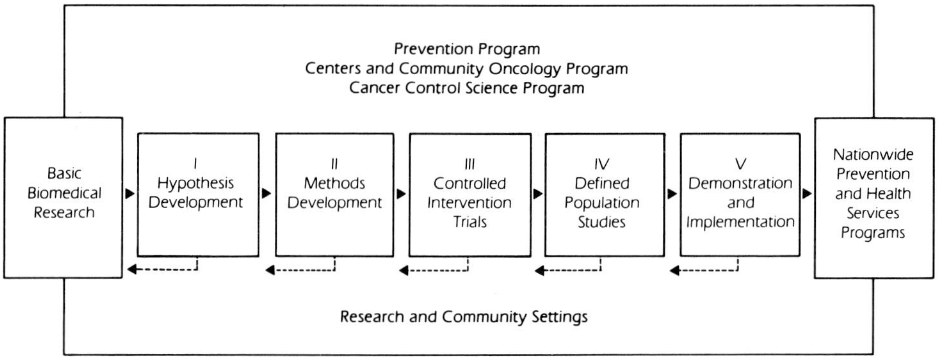

Phase I: *Hypothesis development*. Scientific evidence from laboratory, clinical, epidemiological, or socio-behavioural research is assessed to determine its application. An hypothesis is formulated on the basis of prior scientific research. Hypothesis development is more the identification and synthesis of scientific evidence than the formulation of studies. It provides the framework within which the cancer control interventions are built.

Phase II: *Methods development*. Studies at this level are designed to identify the variables to be controlled or monitored in subsequent interventions and to ensure that accurate and valid procedures are available before an intervention is implemented. The following types of studies fall into this category: the feasibility of using a proposed intervention in a specific population subgroup; developing, pilot-testing and validating data-collection forms, instruments or questionnaires; testing translations of material from other languages; testing, through pilot studies, alternative forms of intervention; testing methods from other medical areas or from other disciplines on cancer problems; developing methods to improve compliance.

Phase III: *Controlled intervention trials*. The aim here is to test the hypotheses developed in Phase I using the methods validated in Phase II. The purpose of these trials is to determine the efficacy of the intervention. The group selected for testing is not necessarily homogeneous with the eventual target population and may be chosen to facilitate research management rather than to provide a representative sample. Interventions may be carried out prospectively as well as examined retrospectively. Randomized-controlled intervention trials are used as the standard to obtain the clearest research outcome. Whatever the experimental or analytical approach taken, the emphasis is on demonstrated efficacy.

Phase IV: *Defined population studies*. Studies in this phase are designed to quantify the impact of an effective intervention in a sample representative of a large target population. Results from these studies, therefore, can be applied to a specific target population.

A defined population may be characterized in terms of demographic characteristics (e.g., age, sex, ethnic group), social and economic factors (e.g., occupation, education, socioeconomic status), vital statistics (e.g., incidence, morbidity, mortality), personal and lifestyle factors (e.g., diet, smoking, sun exposure), genetic and biological characteristics (e.g., individuals with a familial cancer risk or a high-risk biological marker), or other factors associated with cancer or risk factors for cancer. Phase IV studies also resolve new issues that may arise when interventions are applied to population subgroups that are larger than those tested in Phase III studies.

Phases III and IV studies may sometimes be merged in order to have earlier results from an intervention strategy. Since the cost of Phase IV studies is usually much greater than for Phase III, the decision to combine these phases must be determined, at least partially, by a higher probability of a successful outcome.

Phase V: *Demonstration and implementation studies*. The purpose of Phase V studies is to apply interventions that have already proved to be effective. An evaluation scheme as well as quality-control procedures must be defined to ensure that the intervention is carried out using the methodology validated in previous phases. In many cases, these studies may be more cost-effective if they are included in multifaceted programmes (e.g., adding a diet-related intervention to an existing smoking intervention study).

Examples of the need for a new cancer control research strategy

If it is to move from basic research studies to an application in human populations, systematic cancer control research has additional importance. Not only does it guarantee that research does not stagnate at any one stage, it also ensures that appropriate attention is given to safety and ethical issues, and it helps to avoid duplicative, long, and expensive studies. But, as the discussions that follow reveal, the history of cancer control has many examples of application occurring years after the necessary scientific basis was known and available.

Cervical cancer: the Pap test. In 1928, Dr George N. Papanicolaou in the USA and Dr Aurel Babes in Romania reported independently that cancer of the cervix could be diagnosed by examining exfoliated cells from the cervical epithelium (Papanicolaou, 1928; Babes, 1928). Despite the clinical importance of these observations, they were published in rather obscure journals and escaped notice for more than a decade. Papanicolaou, a laboratory scientist, finally devoted all his energy to pursuing his finding when he became associated with a gynaecologist, Dr Herbert F. Traut, who clearly recognized the potential of exfoliative cytology for cancer control (Breslow et al., 1977). Together, the two perfected the breakthrough technology bearing Papanicolaou's name (the Pap test). This methodology permitted practical mass screening for cervical cancer. Their definitive work (Papanicolaou & Traut, 1943) was an early example of the value and necessity of multidisciplinary collaboration in medical science.

More than a decade later, however, the Pap test was only rarely being used. One of the factors causing this delay was the absence of a definitive clinical trial. It was only in the 1950s and early 1960s that the US Government sponsored scattered field studies to demonstrate the feasibility of cervical cytology testing in the community. General adoption of the test finally occurred in the early 1970s, when an estimated 70–90% of women in the USA had been tested at least once (Breslow et al., 1977).

Breast cancer: mammography and physical examination. The under-use of mammography and physical examination to detect breast cancer, the leading cancer killer of women for most of this century, is reminiscent of the delays that characterized the history of cervical cancer control.

As early as 1913, Dr A. Salomon used X-rays to detect the common forms of breast cancer (Salomon, 1913). Seventeen years later, Dr Stafford Warren demonstrated the diagnostic value of breast X-rays (Warren, 1930); and by 1937 Dr Jacob Gershon-Cohen proposed that X-ray screening of asymptomatic women would reduce deaths from breast cancer (Gershon-Cohen & Colcher, 1937). However, nearly a quarter of a century passed before surgeons and radiologists began to test systematically Gershon-Cohen's claims which, by then, had evolved and been modified by other investigators (Breslow et al., 1977). These efforts included reproducibility studies and a randomized clinical trial that began in 1963 supported by the NCI and the Health Insurance Plan (HIP) of the Greater New York Screening Program. Led by Sam Shapiro, HIP's vice-president and director of research and statistics, and Drs Philip Strax and Louis Venet, the study was a model of a well-designed and systematic field evaluation of a screening procedure for cancer (Strax et al., 1970). These studies closely followed a new mammography technique from the early 1960s, that used industrial film that could be reproduced by other radiologists (Egan, 1961).

At this point, the history of the Pap test and mammography studies diverge. Because radiologists were being trained in Egan's technique while the scientific evaluations were still under way, mammography was incorporated into private practice. The potential cancer risk from the radiation dose received at the time became a subject of controversy. Moreover, the positive findings of the HIP study – that periodic screening could significantly reduce breast cancer mortality in women more than 50 years old – influenced the decision for a policy of promoting breast-cancer screening in younger women as well. When the appropriate age for routine application of mammography was debated, confusion resulted among both women and physicians, further slowing the use of mammography and physical examination even in the population of post-menopausal women who were shown in the HIP study to benefit most from these procedures. In the late 1970s, it was necessary to restore confidence in the appropriate use of mammography for women older than 50 years and to emphasize proper evaluation before routinely applying the technique to younger women.

One programme implemented to accomplish these aims was the Breast Cancer Detection Demonstration Project (BCDDP). Funded jointly by the NCI and the ACS, it began in 1973 and screening was completed in 1981 (Baker, 1982). More than 280 000 women were enrolled in 29 centres at 27 widely distributed locations throughout the USA. As a demonstration programme, the BCDDP was not designed to research the effectiveness of screening to reduce mortality. However, the results showed that mammography played an even greater role in the diagnosis of breast cancer than had been indicated in the HIP study in both the 40–49 and 50–59 age groups. This finding was most likely due to the improvements in mammography technology made during the time between the two studies. In addition, in the BCDDP the relative contribution of mammography (compared to that of physical breast examination) to the detection of early cancers was impressively high, emphasizing the value of mammography in early breast-cancer detection. But most importantly, the BCDDP illustrated the need for cause-and-effect intervention research before the commencement of costly demonstration programmes, not vice versa.

These two examples of an unsystematic approach to cancer control involve what many refer to as secondary prevention, that is, the early detection of cancer for the purposes of beginning treatment as early as possible, thereby increasing the chances of survival and reducing mortality. The following discussion relates to similar deficiencies in scientific logic relating to primary prevention, i.e., preventing the occurrence of cancer itself.

Smoking and lung cancer. After decades of informal medical observations, particularly by thoracic surgeons, Drs Ernst Wynder and Evarts Graham (Wynder & Graham, 1950) reported the first epidemiological study that showed a clear association between smoking and lung cancer. At the same time, Dr Morton L. Levin and his colleagues (Levin *et al.*, 1950) also found that their data also suggested a causal relationship between smoking and lung cancer.

From 1950 to the mid-1960s, a number of other studies confirmed conclusively the information showing smoking to be the major cause of lung cancer. Chief among these was a landmark ten-year study of smoking-related lung cancer mortality in 40 000 British physicians conducted by epidemiologist Sir Richard Doll and statistician Dr Austin Bradford Hill (Doll & Hill, 1956). These investigators provided evidence that lung cancer mortality increases in direct relation to the amount a person smokes and that lung cancer incidence would be reduced by 80 to 90% in the absence of smoking. To this study and other investigations was added a signal prospective study in the USA organized by Drs Cuyler Hammond and Daniel Horn of the ACS (Hammond & Horn, 1958). It involved observations of 188 000 men selected by their smoking habits. The results supported the conclusions of the preceding studies and showed, in addition, that smoking increased mortality from heart disease.

The overwhelming amount of epidemiological and laboratory evidence prompted a broader examination of the health implications of tobacco use by an expert committee appointed in 1962 to advise the Surgeon General, Dr Luther Terry. The outcome was the authoritative 1964 report on smoking and health (US Department of Health, Education, and Welfare, 1964) that concluded that cigarette smokers have a higher death rate from heart disease than non-smokers. Subsequent studies have established smoking as a major cause of cancers of the larynx, oral cavity and oesophagus and at least a contributory factor in cancers of the bladder, kidney and pancreas (US Department of Health and Human Services, 1982).

The 1964 and later reports of the Surgeons General on smoking refer to thousands of studies, including eight major prospective studies and more than 50 retrospective investigations, which are impressively consistent in their findings (Cullen, 1986). Unlike the state of the science in early cancer detection, most of the smoking-and-health research became redundant by the early 1970s, providing only diminishing returns on both the resources spent and the scientific knowledge available. The result was that substantial time was lost in initiating effective public health measures to control smoking. Had a national policy existed in the 1960s or by the 1970s, attention might have shifted to the development of technology to reduce smoking prevalence.

Opportunities for the future

Given the above experiences, we must now carefully avoid redundant studies before the application of cancer research. But we must also avoid starting intervention studies prematurely before efficacy and safety have been proven. The need for an organized research strategy is best illustrated today in the area of diet and cancer, where there are many important etiological connections between diet and cancer risk (Gershon-Cohen & Colcher, 1937) but not sufficient evidence to prescribe exactly what can be done to prevent cancer. Unlike tobacco use and cancer, where public health opportunities from clinical trials to large demonstration studies are evident, intervention programmes involving dietary factors are principally focused on clinical trials and efficacy studies with only general prescriptions suitable for demonstrational purposes.

By the 1980s, the NCI was faced with a growing public interest and public demand that the Institute interpret existing dietary information while the etiological and clinical-trial research continued. Accordingly, interim guidelines have been developed and a strong research initiative has begun.

In the early 1980s, the NCI commissioned the National Academy of Sciences to summarize the state of the knowledge on diet and cancer. The Academy's report (NAS, 1982) contained a conservative list of dietary recommendations. These recommendations were similar to those provided by two separate US agencies (US Department of Agriculture, 1980; US Department of Health and Human Services, 1984). These guidelines are consistent with good nutritional practice and following them may reduce the risk of cancer. Both the NCI and the ACS issued their own dietary recommendations in 1984, encouraging people to eat a variety of foods that are high in fibre, low in fat, and sufficiently low in calories to maintain a correct body weight.

The challenge of applying the phases of cancer control research

The execution of each of these phases presents various challenges. *Hypothesis development* (phase I) appears to be the simplest of the phases and the least difficult to accomplish. But the available scientific mechanisms that continuously and systematically review the basic research and epidemiological literature in order to find promising hypotheses to reduce cancer risk and rates are scarce and not well developed. For the most part, phase I hypotheses tend to be generated randomly and go unprioritized in relation to other types of hypotheses also being formulated.

Methods development (phase II) presents a different set of problems. It is often assumed that reliable and valid methods to test hypotheses exist and are adequate. Scientists interested in the relationship between diet and cancer, however, are aware of the fallacies of such assumptions when they have attempted to establish reliable dietary histories or to identify physical or chemical markers of food intake for the purpose of determining dietary compliance or cancer risk.

The challenges for the efficacy phases, *controlled intervention trials* (Phase III) and *defined population studies* (Phase IV) are numerous. Such studies are expensive to support, difficult to implement, and require a long-term commitment to complete. Few scientists know how to carry them out or are interested in doing so. Since the development of the cancer control phases in the early 1980s, over 100 efficacy trials (predominantly Phase III) have now begun in the USA alone. About 60% of these trials are directed towards reducing tobacco-use prevalence, particularly cigarette smoking. The remainder are principally diet-related, most being chemoprevention trials.

While there was a general consensus that it was time to move from etiological studies relating smoking to cancer to developing and testing effective strategies to control smoking, less enthusiasm is evident for diet-related intervention trials. But those supporting the need for such trials defend their propositions by arguing that more epidemiological and experimental evidence will not provide a definitive answer to the question about how to modify dietary behaviour or how to reduce cancer mortality. The real definitive test is to carry out a prospective clinical trial in humans, particularly one where the subjects are randomized to the study conditions.

The Women's Health Trial (WHT), started by the NCI in 1986, is a striking example of the problems facing Phase III intervention trials for diet and cancer studies. The randomized field trial was to be conducted over a ten-year period. It was intended to evaluate the hypothesis that a reduction in the proportion of dietary calories from fat, i.e., from about 40% to half of that amount, would reduce breast cancer incidence by one-half in women at increased risk of the disease. The WHT was designed to involve 32 000 women, aged 45 to 69 at enrolment, drawn from 20 collaborating centres across the USA.

The trial began with a feasibility study, testing whether enough women could be introduced into the trial to satisfy the statistical power calculations and whether these women could comply with the difficult dietary conditions. Although the answer to both those questions was positive following the two-year feasibility study, the trial was discontinued. The decision to terminate the trial resulted from an advisory committee's review that found that the hypothesis was not sufficiently strong, that there were not enough subjects, and that there was a possibility of a false negative result, since the control subjects might reduce their fat consumption in keeping with current trends.

The decision to terminate this trial will continue to be as controversial as the decision to mount it. Such controversies magnify the importance of having a systematic decision-making process to determine whether large and expensive studies ought to be conducted.

Finally, in Phase V, *demonstrations*, the challenges are immense. There are critics who do not consider such studies to be scientifically rigorous. There are also those who argue for more of them, since the potential for a public health benefit is great. Although the evaluation of demonstrations is usually less rigorous compared with that for controlled trials, the necessity to determine whether they are successful is important. Such demonstrations are expensive and labour intensive. At present, there are too few examples of demonstrations and their evaluation to allow for a meaningful review.

The reasoning behind cancer control discussed here is hardly infallible. But it is consistent and it does attempt to provide a framework for thoughtful decision-making. If it is used, cancer control will most likely be improved. Hopefully, using this type of cancer control, cancer risk and cancer rates will be reduced in the populations under study.

Conclusion

As the understanding of the social impact of cancer has advanced, three major actions have shaped the current state-of-the-art in cancer prevention and control: the maturation of organizational approaches to solving the cancer problem; the accumulation of scientific evidence that justified the concept of cancer as a controllable disease; and most recently, the development of a systematic cancer control research process that provides efficient and clearer direction to the planning of a national cancer control programme. The NCI, recognizing the need for goal setting and scientific accountability in the conduct of cancer control research, has established mortality reduction objectives for the year 2000 related to cancer prevention. The cancer control research process and these objectives are guiding the allocation of cancer control resources for the application of effective health promotion strategies.

References

American Cancer Society (1984) Nutrition and cancer: causes and prevention. *Ca-A Cancer Journal for Clinicians*, 34, 121-126

Babes, A. (1928) Diagnosis of cervical cancer by cervical smear [in French]. *Presse Med.*, 36, 451-454

Baker, L.H. (1982) Breast cancer detection demonstration project: five-year summary report. *Ca-A Cancer Journal for Clinicians*, 32, 194-225

Breslow, L., Agran, L., Breslow, D.M., Morganstern, M. & Ellwein, L. (1977) Cancer control: implications from its history. *J. Natl Cancer Inst.*, 59, 671-686

Cullen, J.W. (1986) A rationale for health promotion in cancer control. *Prev. Med.*, 15, 442-450

Doll, R. & Hill, A.B. (1956) Lung cancer and other causes of death in relation to smoking. A second report on the mortality of British doctors. *Br. Med. J.*, 2, 1071-1081

Egan, R. (1961) Letter to the editor. *J. Am. Med. Assoc.*, 178, 440

Gershon-Cohen, J. & Colcher, A.E. (1937) Evaluation of roentgen diagnosis of early carcinoma of the breast. *J. Am. Med. Assoc.*, 108, 867-871

Greenwald, P. & Cullen, J.W. (1984) The scientific approach to cancer control. *Ca-A Cancer Journal for Clinicians*, 34, 329-332

Greenwald, P. & Cullen, J.W. (1985) The new emphasis in cancer control. *J. Natl Cancer Inst.*, 73, 543-551

Greenwald, P. & Sondik, E.J., eds (1986) *Cancer Control Objectives for the Nation: 1985-2000*, NCI Monograph No. 2 (NIH Publication No. 86-2880), Bethesda, MD, US Department of Health and Human Services

Hammond, E.C. & Horn, D. (1958) Smoking and death rates - report on forty-four months of follow-up of 187 783 men. II. Death rates by cause. *J. Am. Med. Assoc.*, 166, 1294-1308

Levin, M.L., Goldstein, H. & Gerhardt, P.R. (1950) Cancer and tobacco smoking. A preliminary report. *J. Am. Med. Assoc.*, 143, 336-338

National Academy of Sciences (1982) Diet, nutrition, and cancer. *Report from the Committee on Diet, Nutrition, and Cancer*. Assembly of Life Sciences, National Research Council, Washington DC, National Academy Press

Papanicolaou, G.N. (1928) New cancer diagnosis. In: *Proceedings of the Third Race Betterment Conference*, January 2-6, 1928, Battle Creek, Michigan, Race Betterment Foundation, 528-534

Papanicolaou, G.N. & Traut, H.F. (1943) *Diagnosis of Uterine Cancer by the Vaginal Smear*. New York, Commonwealth Fund

Salomon, A. (1913) Contributions to the pathology and clinical aspects of mammary carcinoma [in German]. *Arch. Klin. Chir.*, 102, 573-668

Strax, P., Venet, L., Shapiro, S., Gross, S. & Venet, W. (1970) Breast cancer found on repetitive examination in mass screening. *Arch. Environ. Health*, 20, 758-763

US Department of Agriculture and US Department of Health and Human Services (1980) *Nutrition and Your Health: Dietary Guidelines for Americans* (Home and Garden Building No. 232), Washington, DC, US Government Printing Office

US Department of Health, Education and Welfare (1964) *Smoking and Health. Report of the Advisory Committee to the Surgeon General of the Public Health Service*, Washington DC, Center for Disease Control, Public Health Service Publication No. 1103

US Department of Health and Human Services (1982) *Health Consequences of Smoking: Cancer. A Report of the Surgeon General*, Office of Smoking and Health, Rockville, MD, DHHS Publication No. (PHS)82-50179

US Department of Health and Human Services (1984) *Diet, Nutrition, and Cancer Prevention: A Guide to Food Choices*, Bethesda, MD, NIH Publication No. 85-2711

Warren, S.L. (1930) A roentgenologic study of the breast. *Am. J. Roentgenol. Rad. Ther.*, 24, 113-124

Wynder, E.L. & Graham, E.A. (1950) Tobacco smoking as a possible etiologic factor in bronchiogenic carcinoma. A study of six hundred and eighty-four proved cases. *J. Am. Med. Assoc.*, *143*, 329-336

75th Congress. *The National Cancer Institute Act of 1937*. Senate Bill 2067. Public Law 244. Signed August 5, 1937

92nd Congress. The National Cancer Act of 1971. Senate Bill 1828. Public Law 92-218. Signed December 23, 1971

Evaluating Effectiveness of Primary Prevention of Cancer
Ed. M. Hakama, V. Beral, J.W. Cullen & D.M. Parkin
Lyon, International Agency for Research on Cancer
© IARC, 1990

QUANTIFICATION OF THE EFFECTS OF PREVENTIVE MEASURES

J. Kaldor[1] and D.P. Byar[2]

[1]*International Agency for Research on Cancer, Lyon, France*
[2]*National Cancer Institute, Bethesda, MD, USA*

Introduction

Prevention of cancer is achieved when modification of exogenous factors results in a reduction in cancer risk. This definition covers a wide variety of possible preventive measures, from avoidance or reduction of exposure to carcinogenic agents, to increased consumption of micronutrients. Although some authors (Bertram *et al.*, 1987; Meyskens, 1988) have found it conceptually useful to describe prevention activities according to the stages in the process of carcinogenesis at which they are presumed to act, we will not make these distinctions in this article. The number of stages involved in carcinogenesis may vary with the particular carcinogen and the site upon which it acts, and at our current level of understanding it is not generally possible to discern the stages of human carcinogenesis with useful precision. It is likely, however, that further research may permit such distinctions. In any case, when considering quantification of the effects of preventive measures, we may accept that a measure is effective even if we do not understand its mechanism. For example, a change in dietary pattern may also entail modifications in physical activity, weight, and exposure to carcinogens such as alcoholic beverages, any or all of which could lead to a reduction of cancer risk.

Quantification of the effect of prevention can be made at several levels of detail, depending on the intervention strategy under consideration, and the type of data available for its evaluation. At the crudest level, the effect of a preventive measure is quantified by comparison of risk between populations in which it has and has not been applied. At the other end of the spectrum is a full quantitative description of the reduction in risk as a function of the degree to which the measure is applied, the time course of the reduction, and the composition of the population in which it is implemented.

The purpose of this paper is to review some indices which can be used to quantify the effectiveness of preventive measures, and methods for their estimation from randomized and observational studies. We then consider the design of studies to evaluate preventive strategies, making reference to mathematical models of carcinogenesis and evaluation of negative results.

Indices for quantification of primary prevention

Studies in which prevention strategies are evaluated can be viewed as the mirror image of etiological studies: rather than examining factors which increase cancer risk, they focus on the reduction of risk, often by modification of exactly the same factors. Therefore, the same indices can be applied to describe the relationship between a preventive measure and the consequent reduction in cancer risk. Three indices of effectiveness are of fundamental importance. Although they are related, each conveys a somewhat different aspect of the benefit obtained by a preventive measure.

Define P_i to be the probability of cancer occurrence in an individual to whom the intervention has been applied, and P_c the probability in a comparable control individual. The ratio $R = P_c/P_i$ is the relative risk between the two, and $R_1 = 1 - P_i/P_c = 1 - 1/R$ is the proportion of risk which is removed by the intervention. For example, if the intervention reduces risk by 20%, then $R = 1.25$, and $R_1 = 1 - 0.8 = 0.2$. This quantity, which may be referred to as the *proportional risk reduction* is easily interpreted by non-scientists because it represents the average proportion by which someone can expect to reduce their cancer risk by the preventive measure. At the population level, of more interest may be the *absolute risk reduction*, $R_2 = P_c - P_i$, which, when applied to the number of people subject to the preventive measure, gives the number of preventable cases.

A population may consist of two groups, only one of which would benefit from the intervention. For example, smoking cessation has no relevance to cancer risk in non-smokers. Defining f as the proportion of the population in whom the measure can be effectively applied, P_c and P_i as the probabilities of developing cancer in that proportion of the population before and after intervention, and P_b as the background probability of cancer in the remainder of the population (see Figure 1), the *population reducible risk* R_3 can be defined as:

$$R = \frac{f(P_c - P_i)}{fP_c + (1 - f)P_b},$$

the proportion of cancer in the total population which would be removed by the intervention. In this expression, the numerator indicates the reduction in risk experienced by those for whom the intervention has an effect, and the denominator gives the total risk in the population before intervention. If the result of the intervention is to reduce risk to the background level P_b, so that $P_i = P_b$, then the population reducible risk becomes

$$\frac{f(R - 1)}{1 + f(R - 1)}$$

the *population attributable risk* (Levin, 1953) for an exposure with prevalence f. Sturmans et al. (1977) and Deubner et al. (1980) have suggested that in some situations where disease prevention is being considered, it is of interest to investigate the reduction of high levels of specified factors to realistic target levels.

Figure 1. Schematic illustration of the relationship between the probability of cancer occurrence before (P_c) and after intervention (P_i) in the proportion f of the population subject to intervention, and of the background probability of cancer occurrence (P_b) in the proportion $(1-f)$ of the population not subject to the intervention

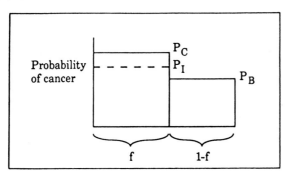

The formulation above assumes that a preventive measure can be applied in exactly the same way to all members of the population subject to intervention. In practice, the measure is implemented to a varying extent in the population under study. For example, a dietary modification such as an increase in fruit consumption or a reduction in fat intake is unlikely to be adopted by all members of the population to the same degree. In this situation, the simple summary indices based on P_i and P_c are inappropriate, since they do not describe the effect of the primary prevention measure in sufficient detail. The indices could be refined to take account of variation in implementation levels, by defining them separately for several categories of modification. For the example of fat intake, one could adopt as indices the risk reduction associated with a decrease of less than 5%, 5–10%, 10–20%, or more than 20% calories as fat per day.

This solution encounters difficulties when there is a wide range in starting values of the factor being modified, as well as a range in the degree of modification, since it then becomes necessary to define the indices for a large number of combinations of the two scales. Wahrendorf (1987) has proposed that the effect of an intervention measure of this kind should be quantified in relation to a mean shift in the distribution of the factor to be modified by the prevention measure. Considering a bivariate distribution of fat consumption in individuals before and after intervention, one might expect a correlation between the two levels, reflecting the likelihood that, on average, a high consumer will remain a high consumer, and a low consumer similarly. The effect of the dietary modification on cancer risk could then still be quantified by the summary indices defined above, but they would refer to the effect of the mean shift, rather than of a constant, quantum change for each individual.

An index of cancer risk reduction must be qualified by the time period required for the indicated effect to take place. Statements such as 'Elimination of smoking would reduce lung cancer incidence by X%' are incomplete unless the time needed for the reduction to take place is also specified. Like the indices themselves, their temporal characteristics have both an individual and a population aspect. For an individual, the time pattern of the risk to be expected following implementation of a preventive measure depends on a number of factors, including age, past exposure history, and the nature of the measure. The effect on lung cancer risk of stopping exposure to asbestos fibres by changes in industrial hygiene is likely to be quite different for a recently employed 30-year-old non-smoker than for a heavy smoker who has spent a full working life in contact with asbestos. At the population level, the temporal pattern of risk reduction depends on the distribution of such determining factors in the ensemble of exposed individuals in the population.

Before leaving indices of effect, a word is required about modification of multiple factors. If the effects of the factors in reduction of cancer risk are multiplicative, they can be described separately because the reduction due to one factor is independent of the level or change of other factors (Rothman, 1976). When the relationships are not multiplicative, the effect of modification of one factor cannot be simply quantified, because it depends on the level of the other factors. For example, removal of an occupational lung carcinogen whose mechanism of action is additive to that of tobacco smoke will have a much more dramatic proportional effect in a population with low smoking prevalence than in one with a high proportion of smokers.

Estimation of primary prevention indices

For obvious ethical reasons, human etiological studies can never be randomized. Randomized prevention studies also run into opposition on the grounds that an intervention measure which has received substantial support from observational or experimental data should not be systematically withheld, as it would be from the control group in a randomized study. However, sufficient uncertainty has prevailed in several cases to allow randomization to be incorporated into cancer prevention studies. Randomized studies provide the most straightforward means of estimating the effect of intervention, because they are (at least theoretically) immune to confounding, the fundamental problem of observational epidemiology. The effect measures R_1 and R_2 defined above can be estimated directly and without bias making direct use of the estimates of P_i and P_c from the two arms of a study. Estimation of the population reducible risk R_3 requires some further data, possibly from outside the prevention trial, on the proportion f and the probability P_b of cancer in those unaffected by intervention. In interpreting the results, any observed reduction in cancer risk can be attributed to the intervention, as long as the randomization has been carried out adequately. One limitation of randomized trials is that they are often carried out, for good logistical and statistical reasons, on rather homogeneous populations, and the generalizability of results to other groups is not assured. For example, a Finnish lung cancer prevention trial is being carried out only on the subgroup defined by male smokers aged 40–69 (Huttunen, 1985).

Randomized trials remain the exception in the evaluation of preventive strategies, however, and one must rely on observational data, sometimes arising from 'natural experiments' on risk factor modifications. Observational data may come from cohort studies, which involve estimation of the risk of cancer for people who have been followed up for long time periods, or from case–control studies, in which cancer cases are compared with control subjects. Another possible source of data on the effect of intervention is the ecological study, in which the correlation between disease risk and the factors defining the intervention is calculated across population units. All three study designs permit the estimation of relative risk, but the estimation of f, the proportion of the population susceptible to intervention, requires a random sample of the population of interest. A sample of this kind is provided by a case–control study if the controls are randomly drawn from the population, or a cohort study whose subjects are a random sample of that population. In fact, the control group in case–control studies is often a random sample of a population which has been matched to the cases by age and sex, but cohort studies consisting of random samples of defined populations are somewhat less common. More often the cohort is in some way selected, for example consisting of a group of volunteers, workers in a specific industry or members of a professional organization.

The effect of a particular intervention is estimated by subdividing the study population into those who have and have not undergone the intervention, and estimating the relative risk between the two groups. The estimation of effect from observational studies is complicated by the need to take account of confounding. There is no way to exclude the possibility that observed effects are due to factors other than those being used to characterize the intervention, or to ensure that the effect would have occurred if the intervention had been applied at random to one group. For example, inferences from a cohort study in which those who give up drinking spirits have a lower risk of oral cancer than continuing drinkers can always be challenged on the grounds that cessation of spirit consumption may have been accompanied by other changes, such as reduction in smoking. The standard epidemiological solution to this difficulty is to adjust for other factors in the statistical analyses, but adjustment for a variable requires that it has been recognized or suspected as relevant in the etiology of the disease, and included in the study questionnaire. Bruzzi *et al.* (1985) show how the population attributable risk for multiple factors can be estimated, with or without adjustment for other (confounding) factors, using data from case–control studies analysed by logistic regression. They also explain how the same model can be used to assess the effects of preventive strategies.

Because of the potential for unadjusted confounding, the magnitude of risk reduction estimates obtained from observational studies may be determined by factors other than the one whose effect is being measured and they should therefore be viewed with caution. For example, a study of a vegetarian population, such as the Seventh-Day Adventists, may yield a breast cancer risk which is 30% lower than that of a non-vegetarian population, even after adjusting for differences in reproductive and other factors related to breast cancer. However, because there may be other differences which distinguish Seventh-Day Adventists from the general population, the 30% might be considered as an upper limit of the effect of dietary modification if we assume that the other factors are most likely ones which would decrease the risk of breast cancer.

Whether arising from randomized or observational studies, estimates of risk reduction should as far as possible be accompanied by standard error estimates. However, until recently (Benichou & Gail, 1989a,b) methodology for estimating these standard errors has not been routinely available except in simple situations (Walter, 1975). Independent estimates of risk reduction provided by several studies can be combined by weighting them inversely to their variances.

Design of prevention studies

The general issues of design for prevention studies are similar to those arising in treatment trials, if the study is randomized, and to etiological investigations, if the study is observational. Decisions must be made concerning the sampling mechanisms for study subjects, the degree of matching employed, the amount of information to be requested, and the numbers of individuals to include.

Randomized trials of prevention using cancer incidence or mortality as outcome measures are likely to be substantially bigger and more costly than therapeutic trials or even prevention trials in which premalignant lesions are the outcomes of interest. Even in high risk populations, only a relatively small number of study subjects will develop cancer over a 5–10-year period and large numbers of subjects will generally be required to study prevention. Byar and Piantadosi (1985) and Byar (1988) have argued that considerable cost savings can be obtained using factorial designs to study several interventions simultaneously. More recent design features suitable for prevention trials but seldom used in treatment trials include group randomization and case–cohort monitoring schemes (Prentice, 1986), both of which can reduce costs considerably. Byar (1988) gives some examples of these strategies.

If a randomized trial is to be implemented, a substantial amount of observational data has probably already accumulated in support of the proposed intervention. However, these data may give no indication of the magnitude of the possible effect to be expected or the time required for it to take place. For example, intervention aimed at dietary factors is often justified mainly by correlation studies which show that average levels of nutrients are correlated with cancer risk across regions or countries (Armstrong & Doll, 1975). Recently proposed or implemented chemoprevention studies for oesophageal cancer (Erchow et al., 1984) and breast cancer (Greenwald et al., 1987) have been largely premised on observations of this kind. Although ecological correlation studies may provide useful estimates of the possible magnitude of effects to be expected, they provide no information on the age at which intervention should begin, how long it must be continued, or how long one must wait to see maximal effects. Studies demonstrating a change in risk with migration between countries or regions of differing risk levels can help in answering these questions, especially when information on changes in risk factor levels is available.

Randomized trials have been implemented to study the reduction in primary liver cancer risk resulting from vaccination against the hepatitis B virus (HBV). The effect of HBV vaccination can be predicted with greater accuracy than that of dietary modifications for several reasons. Firstly, it is known that the vaccine is only effective in endemic regions when administered soon after birth, so that the target population for intervention is very clearly defined. Secondly, the relative risk associated with the HBV carrier status is so large that it is unlikely to be explainable by confounding (Schlesselman, 1978). Finally, the intended effect of the intervention is the elimination of the carrier status and the enormous increase in risk which it conveys (Beasley *et al.*, 1981), as compared to the rather modest modifications which are expected to occur in some dietary intervention studies (Prentice *et al.*, 1988). Where so much information is available on the likely consequences of intervention, the ethical difficulties may require that some modified form of randomization be employed. For example, the 'stepped wedge' design used in the ongoing HBV intervention trial in the Gambia (The Gambia Hepatitis Study Group, 1987) involved random assignment of the order in which different areas of the country were to be covered by vaccination, even though the whole country was eventually covered.

Randomization is clearly not indicated when removal of a carcinogen can be accomplished by direct means (as opposed to the indirect pathway of, say, vaccination, with its attendant uncertainties in safety and efficacy). Thus the risk reduction produced by elimination of an occupational carcinogen can only be estimated from an observational study. The study design then involves identification of an appropriate population in whom exposure has ceased or at least been substantially reduced, and a means to evaluate the consequent change in cancer risk.

Whether a study of a preventive measure is randomized or observational, its design must involve some form of power calculation. For a randomized study, the question addressed by the calculation is the number of subjects required to ensure a high probability that a specified risk reduction will be detected as significant, should it take place. For some observational studies, the number of subjects may be limited by whatever 'natural experiment' has occurred, and the power calculation then provides information on the amount of risk reduction which has a high probability of being detected as significant with the given number. In both cases, the calculation becomes important in the interpretation of negative study results, by providing information on the maximum size of effect which could have been missed by the study.

The rapidity with which risk reduction takes place following intervention can be described by using the multistage theory of carcinogenesis (Armitage & Doll, 1954), as a framework for distinguishing carcinogens according to the stage in the progression to malignancy at which they exert their effect. If a planned intervention involves removal of an exposure whose principal effect is to increase the rate at which an early transition towards malignancy takes place, one would predict that the risk reduction will take a long time to appear. If exposure to a late-stage carcinogen is reduced, the effect should be detectable rather more quickly. The excess risks of mesothelioma among asbestos workers (Peto *et al.*, 1982) and of solid tumours due to ionizing radiation exposure (Smith, 1985), continue to increase following cessation of exposure. In contrast, lung cancer risk appears to fall following cessation of smoking when comparison is made to continuing smokers (Doll, 1971) and the increased risk of endometrial cancer following post-menopausal hormone replacement diminishes when the therapy is withdrawn (Jick et al., 1979).

Table 1, drawn from Day and Brown (1980) shows the excess lifetime cancer risk following cessation of exposure to agents acting on the first (early) and penultimate (late) stages in a five-stage model of carcinogenesis. The excess is expressed as a percentage of the lifetime risk which would prevail if exposure were not stopped, and is given for various ages at first exposure and durations of exposure. Consider an individual who begins exposure at age 20. For early-stage agents, the lifetime excess risk remains high (96% of the risk of individuals who continue exposure) if exposure lasts 30 years, even if it ceases at age 50, but for the late-stage carcinogen, a substantially lower excess risk (39%) can be obtained by cessation.

These considerations must be kept in mind in the evaluation of data on risk reduction following intervention. Detection of the effect of removing a late-stage agent does not need long follow-up, and resources can therefore be put into maximizing the number of subjects to be followed. On the other hand, the effect of elimination of exposure to an early-stage carcinogen would never be seen in a study with short follow-up, no matter how many subjects were included. A more efficient design might involve reducing the number of subjects in favour of a longer duration of follow-up.

Table 1. Percentage of excess lifetime risk resulting from stopping exposure to an early- or late-stage carcinogen (from Day & Brown, 1980)

Age started exposure	Birth		20		40	
Stage affected by carcinogen	Early	Late	Early	Late	Early	Late
Duration of exposure (years)						
2	12	<1	15	1	20	4
5	28	<1	35	2	44	11
10	50	<1	58	6	71	25
20	77	2	85	19	94	54
30	90	8	96	39	99	79
40	97	21	99	62	>99	94
Lifetime	100	100	100	100	100	100

Conclusion

The evaluation of the effects of preventive measures involves statistical techniques which closely parallel those used in designing and analysing clinical trials and epidemiological studies. Of prime importance are the definition and accurate estimation of indices of effectiveness, which should take account of the composition of the study population, the time course of the risk reduction, and possible confounding if the study is not randomized.

References

Armitage, P. & Doll, R. (1954) The age distribution of cancer and a multi-stage theory of carcinogenesis. *Br. J. Cancer, 8*, 1-12

Armstrong, B. & Doll, R. (1975) Environmental factors and cancer incidence and mortality in different countries, with special reference to dietary practices. *Int. J. Cancer, 15*, 617-631

Beasley, R.P., Lin, C., Hwang, L.Y. & Chien, C.S. (1981) Hepatocellular carcinoma and hepatitis B virus. A prospective study of 32 707 men in Taiwan. *Lancet, ii*, 1129-1133

Benichou, J. & Gail, M.H. (1989a) Variance calculations and confidence intervals for estimates of the attributable risk based on logistic models. *Biometrics* (in press)

Benichou, J. & Gail, M.H. (1989b) A 'delta method' for implicitly defined random variables. *American Statistician* (in press)

Bertram, J.S., Kolonel, L.N. & Meyskens, F.L. (1987) Rationale and strategies for chemoprevention of cancer in humans. *Cancer Res., 47*, 3012-3031

Bruzzi, P., Green, S.B., Byar, D.P., Brinton, L.A. & Schairer, C. (1985) Estimating the population attributable risk for multiple risk factors using case-control data. *Am. J. Epidemiol., 122*, 904-914

Byar, D.P. (1988) The design of cancer prevention trials. In: Scheurlen, H., Kay, R. & Baum, M., eds, *Cancer Clinical Trials: A Critical Appraisal* (Recent Results in Cancer Research, Vol. 111), Berlin, Heidelberg, New York, Springer-Verlag, pp. 34-48

Byar, D.P. & Piantadosi, S. (1985) Factorial designs for randomized trials. *Cancer Treatment Reports, 69*, 1055-1063

Day, N.E. & Brown, C.C. (1980) Multistage models and primary prevention of cancer. *J. Natl Cancer Inst., 64*, 977-989

Deubner, D.C., Wilkinson, W.E., Helms, M.J., Tyroler, H.A. & Hames, C.G. (1980) Logistic model estimation of death attributable to risk factors for cardiovascular disease in Evans County, Georgia. *Am. J. Epidemiol., 112*, 135-143

Doll, R. (1971) The age distribution of cancer: implications for models of carcinogenesis. *J. R. Stat. Soc. A, 134*, 133-156

Erschow, A.G., Zheng, S.-F., Li, G., Li, J., Yang, C.S. & Blot, W.J. (1984) Compliance and nutritional status during feasibility study for an intervention trial in China. *J. Natl Cancer Inst., 73*, 1477-1481

The Gambia Hepatitis Study Group (1987) The Gambia Hepatitis Intervention Study. *Cancer Res., 47*, 5782-5787

Greenwald, P., Cullen, J.W. & McKenna, J.W. (1987) Cancer prevention and control: from research through applications. *J. Natl Cancer Inst., 79*, 389-400

Huttunen, J.K. (1985) Studies on diet, nutrition and cancer in Finland. In: Joossens, J.V., Hill, M. & Geboers, J., eds, *Diet and Human Carcinogenesis*, Amsterdam, Elsevier, pp. 199-206

Jick, H., Watkins, R.N., Hunter, J.R., Dinan, B.J., Madsen, S., Rothman, K.J. & Walker, A.M. (1979) Replacement estrogens and endometrial cancer. *New Engl. J. Med., 300*, 218-222

Levin, M.L. (1953) The occurrence of lung cancer in man. *Acta Unio. Int. Cancer, 9*, 531-541

Meyskens, F.L. (1988) Thinking about cancer causality and chemoprevention. *J. Natl Cancer Inst., 80*, 1278-1281

Peto, J., Seidman, H. & Selikoff, I.J. (1982) Mesothelioma mortality in asbestos workers: implications for models of carcinogenesis and risk assessment. *Br. J. Cancer, 45*, 124-135

Prentice, R.L. (1986) A case-cohort design for epidemiologic cohort studies and disease prevention trials. *Biometrika, 73*, 1-11

Prentice, R.L., Kakar, F., Hursting, S., Sheppard, L., Klein, R. & Kushi, L.H. (1988) Review: Aspects of the rationale for the women's health trial. *J. Natl Cancer Inst., 80*, 802-814

Rothman, K. (1976) The estimation of synergy or antagonism. *Am. J. Epidemiol., 103*, 506-511

Schlesselman, J.J. (1978) Assessing effects of confounding variables. *Am. J. Epidemiol., 100*, 3-9

Smith, P.G. (1985) Radiation. In: Vessey, M.P. & Gray, M., eds, *Cancer Risks and Prevention*, Oxford, Oxford University Press, pp. 119-148

Sturmans, F., Mulder, P.G.H. & Valkenburg, H.A. (1977) Estimation of the possible effect of interventive measures in the area of ischemic heart diseases by the attributable risk percentage. *Am. J. Epidemiol.*, *105*, 281-289

Wahrendorf, J. (1987) An estimate of the proportion of colo-rectal and stomach cancers which might be prevented by certain changes in dietary habits. *Int. J. Cancer*, *40*, 625-628

Walter, S.D. (1975) The distribution of Levin's measure of attributable risk. *Biometrika*, *62*, 371-374

Evaluating Effectiveness of Primary Prevention of Cancer
Ed. M. Hakama, V. Beral, J.W. Cullen & D.M. Parkin
Lyon, International Agency for Research on Cancer
© IARC, 1990

EFFECTIVENESS OF PRIMARY PREVENTION OF OCCUPATIONAL EXPOSURES ON CANCER RISK

A.J. Swerdlow

*Department of Epidemiology and Population Sciences,
London School of Hygiene and Tropical Medicine,
London, UK*

Introduction

Successful prevention is both the main purpose of etiological epidemiology and, almost paradoxically, a key element in evidence of causation. Since a large proportion of known carcinogens, particularly those which have long been known, are occupational (Doll & Peto, 1981), it might be expected that the literature on occupational epidemiology would be replete with instances of carcinogenic exposures identified and then eliminated, with reduced risk of cancer ensuing. In fact there are not many examples where studies have given clear-cut evidence of this, although one assumes that it must have occurred frequently.

To discuss this further, it is necessary to consider what would count as evidence of effective prevention. In a liberal interpretation, dose–response and duration–response effects might be taken as evidence of effectiveness of prevention, since those workers who were 'prevented' from undergoing intense or long exposures had lower risks than those who were not so 'prevented'. For the present purpose, however, I will take more restrictive criteria – that either there has been a secular sequence of high levels of exposure and high levels of risk followed by lower exposure with lower risk demonstrated, or that risk in a group of workers is clearly much lower than would be expected from their potential exposures, and this reduction can be attributed to a preventive measure. I will not, however, take it as necessary that the diminution of exposure was deliberately intended to prevent the specific cancer; firstly, because, in assessing the efficacy of prevention and in considering prevention as evidence of etiology, the strength of evidence from an epidemiological viewpoint is not dependent on intentions, and secondly, because it is often unclear from the literature what specific motives underlay particular decreases in exposure over time.

Evidence for prevention

Prevention of cancer in relation to occupation may occur in two ways – membership of an occupational group may lead to exposures or behaviour patterns which prevent cancer, or an occupational exposure or behaviour which is carcinogenic may be prevented. The latter type of prevention is the main focus of this paper, but it is worth noting in the context of this volume overall that the former also gives some impressive examples of effective prevention: for instance the greatly reduced risk of cervix cancer in nuns (Fraumeni et al., 1969; Kinlen, 1982) (i.e., the prevention of cervix cancer in nuns by their abstinence from sexual intercourse), and the substantial decrease in lung cancer mortality in doctors relative to the general population in England and Wales, as doctors have reduced their smoking more than the general population has (Doll & Peto, 1976). Perhaps also, the relatively low lung cancer rates in some groups of miners may reflect the prohibition on their smoking at work.

To return to the issue of prevention of occupationally-induced cancers, discussion here will be divided according to the type of preventive measure and the type of evidence involved. All the evidence discussed is from observational studies; there does not appear to be any material in this field from preventive trials, nor in general would they be a realistic approach.

Non-introduction of a carcinogenic agent

One of the most successful forms of prevention is to detect carcinogenicity of a substance in advance, and as a result never to introduce it to the workplace. If the substance is not used anywhere, evidence of efficacy of prevention depends not on epidemiology, but on the validity of the laboratory tests used to ascertain potential carcinogenicity. If non-introduction is less absolute, however – if for instance it is limited to some countries – then epidemiological evidence becomes relevant. Thus, 4-aminobiphenyl (xenylamine), a potent bladder carcinogen when used in the USA (Melick et al., 1971), was never manufactured commercially in the UK nor used on any great scale (G. Parkes, unpublished) because of concern that it would prove to be carcinogenic. Hence there must have been virtually complete prevention of cancer from this agent in the UK, for which no epidemiological evidence of reduced risk can be presented, but none is needed.

Lack of substantial raised risk in persons working with known carcinogens

Another successful form of prevention for which evidence of effectiveness is indirect is where protection of a workforce has been sufficient to ensure that appreciable increased risk has never occurred. Thus, data on workers in the nuclear industry have not so far identified any major risk of cancer, and the data overall suggest that if there is any risk it is very small; only for myeloma is there considerably increased mortality in the most highly exposed workers (Beral et al., 1987).

A more historic instance in which successful preventive measures in some countries may have avoided substantial risk is suggested by the geographic restriction of the association of scrotal cancer with the occupation of chimney sweep. The association was first described and was strong in England, but scrotal cancer in sweeps was rare in continental Europe, where hygiene and protective clothing had long been emphasized (Waldron, 1983). Even in Scotland sweeps rarely succumbed to this cancer unless they had worked in a large English town (Syme cited by Doll, 1975). Thus, there is suggestive evidence that some measure outside England (perhaps hygiene, perhaps the method used to clean chimneys, perhaps that the English practice of sifting soot in order to sell it was not followed in other countries) largely prevented scrotal cancer. Results from a recent cohort study in Sweden confirm a low risk of scrotal cancer in sweeps in that country: no deaths occurred from scrotal cancer in 717 deaths (Gustavsson et al., 1987), and no scrotal cancers were seen among 214 incident cancers observed (Gustavsson et al., 1988) in a cohort of over 5000 sweeps. It is of note that since the start of this century Swedish chimney sweeps have had the contractual right to take a bath during working hours at the end of each day. Whether scrotal cancer has now successfully been prevented to the same extent in England is unclear.

Disappearance or abolition of an occupation where carcinogenic exposures have occurred

Another instance of successful prevention where there is limited scope and need for epidemiological evidence of effectiveness, is where an occupation has disappeared or been abolished. Thus, mineral oil exposure of the scrotum in mule-spinners in cotton mills in England led to raised risks of scrotal cancer, but such work virtually ceased soon after the second world war (Waterhouse, 1972), eliminating this source of exposure.

Workers in a Japanese factory manufacturing mustard gas in 1929–45 experienced 33 deaths from respiratory tract neoplasms up to the end of 1967 instead of 0.9 expected deaths (Wada et al., 1968). In a more recent follow-up, without expected numbers stated, 74 out of 675 workers who manufactured the gas and 49 of 598 maintenance workers at the factory had developed respiratory tract cancer (Miller, 1988). In a cohort who had worked in mustard gas manufacture in England during 1938–44, there was a large significant risk of upper respiratory cancer and some evidence that lung cancer too had been caused by the exposure (Easton et al., 1988). The manufacture of poison gas at these factories ceased around the end of the war, and presumably no such risk exists in countries which no longer make chemical weapons. Similarly, gas mask manufacture was a major cause of asbestos exposure in women in the UK (Jones et al., 1988a), but was a wartime activity and ceased thereafter. More recently, work in gas manufacture by coal carbonization, which had been shown to be a risk factor for cancers of the lung, bladder and scrotum (Doll et al., 1972), ceased in mainland UK because all gas works were closed with the advent of natural gas supplies from the North Sea.

Reducing exposure to radon-daughters in mines is particularly difficult to achieve, since personal protection is not practical and provision of sufficient ventilation may be difficult or impossible (Radford & St Clair Renard, 1984), and hence disease prevention by closure of the facility may be particularly important for this type of hazard. In the mines of Schneeberg and Joachimsthal, primary cancer of the lung at one time accounted for half of all deaths in miners, but employment at Schneeberg was reduced from 700–800 men in 1879 to 70 in 1939 (however several hundred men were still employed at Joachimsthal by then) (Lorenz, 1944). Uranium mining in Ontario is another more recent example where the effects of the hazard were greatly decreased by reduction in size of the workforce: in 1959, about 11 000 men were employed in these mines, but by 1965 there were under 2000 (Chovil, 1981). Similarly, the hazards to workers in Britain making α-naphthylamine and benzidine (bladder carcinogens) were prevented by closure of the plants concerned (Lancet, 1965).

Evidence of effectiveness of prevention in such instances could be obtained if workers from these industries were followed to ascertain the effect of cessation of employment in the industry, but I can find no example where the effects of abolition of an industry on risk have been demonstrated in this fashion.

Decreased exposure of workers who leave an occupation, leading to reduced relative risk in these workers over time

Individuals may reduce their exposure by leaving an occupation in several ways – by transfer to another occupation, early termination of all work, or retirement at the usual age. Interpretation of risk in individuals who leave an occupation early is extremely difficult. The reasons (including ill-health) for leaving an industry early, whilst others remain within it, and the types of individuals who choose to leave, render those with early termination biased with respect to future mortality (Fox & Collier, 1976; Delzell & Monson, 1981; Bond *et al.*, 1985), and sometimes also with regard to exposure in the occupation (Enterline & Marsh, 1982a), and with respect to previous occupation and non-occupational exposures (Gubéran & Usel, 1987).

If, however, early termination of employment in a workforce is not selective, useful information on prevention can be gained. Thus follow-up of 820 asbestos workers in New Jersey, most of whom were employed for short periods in the second world war but were not in general career asbestos workers, has shown generally diminishing observed to expected ratios for lung cancer from 20 years onwards after first employment (Table 1) (Walker, 1984) – i.e., prevention of further exposure to asbestos after leaving the industry has apparently led in due course to a decreasing risk of lung cancer.

The difficulties of selection bias in those who leave employment early can also be avoided by following risk in an entire working cohort beyond the age of retirement. Thus the declining risk of lung cancer for asbestos workers more than 35 years from initial exposure, in several cohorts (illustrated for one cohort in Table 1), is most plausibly explained as a reflection of the long-term preventive efficacy of abolition of exposure at retirement (and improved conditions over time) (Walker, 1984). The same interpretation can be made of some other data on asbestos workers which more directly concern retirees: in 1074 retirees (but including some who retired before age 65) from a US asbestos products company, risk of respiratory cancer reached a peak 10–15 years after usual retirement age (i.e., age 65) and then diminished in those aged 80 years and over (but based on small numbers in the oldest age groups) (Enterline *et al.*, 1987). A similar effect, for which the same explanation may apply, has been found in three cohorts of uranium miners (Radford & St Clair Renard, 1984; Howe *et al.*, 1986; Howe *et al.*, 1987) in whom there was raised risk of lung cancer: this relative risk, or the increase in relative risk per working-level month, diminished substantially at ages ≥ 70 years compared to that in younger age groups (although based on fairly

Table 1. Lung cancer mortality in asbestos-exposed workers in relation to duration since first employment.

(*a*) Mortality in male New Jersey amosite asbestos factory workers; mainly short-term employees during the second world war, not career asbestos workers. Followed to end of 1976. Source: Walker (1984)

Years since first employment in 1941–45	Lung cancer[a]	
	o/e	obs.
5–9	1.19	2
10–14	4.97	13
15–19	6.17	21
20–24	3.81	16
25–29	4.39	22
30–34	3.82	19

(*b*) Mortality in male North-American insulation workers. Followed from beginning of 1967 to end of 1976. Source: Selikoff et al. (1980)

Years since first employment	Lung cancer[b]	
	o/e	obs.
< 10	–[c]	0
10–14	1.82	5
15–19	3.17	27
20–24	3.36	57
25–29	4.58	96
30–34	5.59	103
35–39	4.98	57
40–44	3.82	31
≥ 45	2.98	53

[a] Observed numbers of lung cancers based on review of all available evidence, expectations on routine death certificate data.
[b] Observed and expected numbers of lung cancers based on death certificate underlying causes. A similar pattern was present when observed lung cancers were based on 'best evidence', but with observed numbers and o/e ratios a little higher throughout from 10–14 years since first employment onwards.
[c] 0.7 expected
Statistical testing not published for these data.

small numbers in the ≥ 70 age group in each study). Of course there are severe limitations to this approach – data on cause of death (and comparison cause data for the general population from routine statistics) are likely to be inaccurate for the very elderly; comparison data for those aged ≥ 85 years are often not available by age group in routine statistics; undetected losses to follow-up or deaths will inflate person-years particularly at oldest ages; decreasing risks with age could reflect early deaths of the most susceptible individuals; and if the induction period of a tumour is long, few of the cohort may survive to give data on risk at the ages at which decreased risk would be anticipated. Furthermore, whether a decrease in relative risk with age or duration since first exposure gives evidence of effective prevention depends on the type of effect (e.g., additive, multiplicative) that the occupational exposure has on the underlying risk in the general population. This is discussed and illustrated later.

Risk in relation to termination of employment may give particularly useful evidence on prevention when the induction period is short. Enterline and Marsh (1982a) found that Standardized Mortality Ratios (SMRs) for respiratory cancer in a cohort of copper smelter workers exposed to arsenic tended to be highest in the decade immediately after termination of exposure and to decline thereafter. Although presumably this might reflect, in part at least, selectivity in early termination (e.g., because of illness), it would also appear compatible with a short induction-period effect of arsenic and hence with a fairly rapid preventive efficacy of leaving the industry.

Decreased exposure levels in a workplace over time leading to secular decreases in relative risks within a cohort

If exposure levels are reduced in a workplace, through changes in process (e.g., closed systems) or personal protection, the evolution of risk over time in individuals who had been exposed before the change but later benefited from reduced exposure, could give evidence on the effectiveness of prevention.

Women first employed early in this century in the radium dial-painting industry in the United States suffered greatly increased risks of mortality from bone cancer and from leukaemia and blood diseases (Polednak *et al.*, 1978). Radium doses decreased substantially from 1925–26, as detailed below, and follow-up until 1975 of a cohort of 1915–29 entrants (Table 2) demonstrated a major diminution in risk for bone cancer and for leukaemia and blood diseases over time.

Maher and DeFonso (1987) studied respiratory cancer risk in production workers at a chemical plant using chloromethyl ethers, where some decrease in exposure was judged to have occurred from 1948–61, and progressive improvements resulting in further reduction in exposures started in 1962, culminating in the introduction of a closed system in 1971, when exposure was reduced to 'essentially no exposure'. A cohort of men who had worked in chloromethyl ether-exposed jobs at the plant during 1948–July 71, before introduction of the closed system, had a significant three-fold raised risk of respiratory cancer mortality. The risk was greatest during 1960–64 and then decreased, especially after 1974 (Table 2). This decrease was the result of a rise in the expected numbers of cases in succeeding quinquennia, but not a decline in the observed numbers. A sample survey suggested that a high proportion of workers smoked in 1963, and possibly also later.

Sandström *et al.* (1989) analysed cancer incidence in men who had worked at a Swedish copper smelter which had used local ore with a high arsenic content, and also had smelted nickel, with severe hygiene problems, during the mid-1940s.

Table 2. Cancer mortality over time in radium dial-painting industry, chloromethyl ether workers and nickel refinery workers

(a) Mortality of women first employed 1915-29 in US radium dial-painting industry, by calendar time period; followed to end of 1975. Source: Polednak et al. (1978)

Calendar years	Bone cancer[a]		Leukaemia & blood diseases	
	o/e	obs.	o/e	obs.
<1945	212.50***	17	8.06**	5
1945-64	101.68***	10	2.35	2
1965-75	13.22	1	–[b]	0

(b) Respiratory cancer mortality in men who had been in jobs rated as exposed to chloromethyl ethers, by calendar time period; followed to end of 1981. Source: Maher & De Fonso (1987)

Calendar years	Respiratory cancer	
	o/e	obs.
1948-59	–[c]	0
1960-64	5.28**	5
1965-69	4.12**	7
1970-74	3.83**	10
1975-81	1.80	10

(c) Mortality of male nickel refinery workers, Clydach, S. Wales, first employed <1925, by time since first employment. Followed to end of 1981. < 20 years since first employment taken as the baseline for risk estimation and significance testing. Source: data in Kaldor et al. (1986)

Years since first employment	Adjusted analyses[d]						Unadjusted analyses[e]	
	Lung cancer			Nose and nasal sinus cancer			Lung cancer	Nose and nasal sinus cancer
	o/e	obs.	EMR[f]	o/e	obs.	EMR[f]	o/e	o/e
<20	1.0	6	1.0	1.0	1	1.0	1.0	1.0
20-29	0.74	35	2.8*	2.8	19	4.9	1.02	4.75
30-39	0.47	55	4.9**	2.4	17	6.6*	0.67	2.83
40-49	0.22*	31	4.8**	3.1	13	13.9**	0.31	2.17
≥ 50	0.11*	10	2.4	3.4	6	24.7**	0.14	2.00

* $p < 0.05$; ** $p < 0.01$; *** $p < 0.001$
[a] Includes cases with underlying cause coded to radiation accidents and sequelae, but bone cancer mentioned and would have been underlying cause if radiation exposure had not been on the certificate
[b] 0.86 expected
[c] 0.64 expected
[d] Adjusted for age at first employment, year of first employment and an exposure index (duration of work in high risk areas)
[e] Statistical testing not published
[f] [observed − expected]/person-years. (A measure of absolute excess risk)

Risk of lung cancer incidence was raised over two-fold overall in a cohort of workers who were first employed 1928–66 and who were followed-up from 1958 (or first employment, if later) to the end of 1982. The rate of lung cancer in the cohort over time decreased by about half, however, both absolutely and relative to appropriate reference populations, from the late 1970s onwards. A large range of hygiene improvements had been noted, including the appointment of a safety and health committee, an industrial physician and a safety engineer in 1932, improvements in ventilation at the roaster departments during the 1940s, a new gas purification system in these departments in 1953, and further improvements there in 1975 and 1978, with the roasters eventually being replaced in 1980. In the arsenic refinery departments, substantial improvements in methods occurred after 1935, but unsatisfactory methods continued to be used until 1957. Packing of refined arsenic was unsatisfactory until 1962; and arsenic metal plant improvements occurred in the 1970s.

Enterline (1974) showed a major diminution in the raised risk of respiratory cancer over time in a cohort of workers from three US chromate-producing plants who were in employment at the plants during 1937–40 (their date of first employment was not known) and who were followed-up from 1941–60. Observed-to-expected ratios fell from 29.09 (obs. = 16) in 1941–45 to 4.75 (obs. = 15) in 1956–60. This reduction was a result of a fairly constant observed rate (and number of cases) of respiratory cancer mortality in succeeding quinquennia, despite ageing of the cohort, while expected numbers rose. The results would be compatible with substantial reduction in exposure over time. No data were available on dust levels at the plants (Taylor, 1966; Enterline, 1974). It was noted, however, that in the industry generally there were not great changes in methods and equipment during the 1930s and 1940s, but during the 1950s a transition towards closed automated production systems began (Taylor, 1966). Also, duration of stay of workers in the industry may have altered over time (Taylor, 1966).

Data from several studies suggest a relatively rapid reduction in risk of leukaemia after decreases in benzene exposure. Paci *et al.* (1989 and unpublished) assessed leukaemia mortality in workers in shoe manufacturing in Florence. Glues containing benzene (probably over 70% by weight) were introduced in 1953 and there was high exposure to benzene from 1953 to 1962. In 1961–62, improved ventilation systems were installed at the plant. In 1963, the product formula was changed to contain less than 2% benzene; use of the old products then declined, and from 1965 benzene levels were very low. Observed/expected ratios for leukaemia and aplastic anaemia in men exposed during 1953–64 (i.e. the era of high and intermediate levels of benzene exposure) who were followed up to the end of 1984 were around 10 to 15-fold raised in follow-up periods under 20 years from first exposure (based on 11 cases overall), but in comparison decreased several-fold for the period ≥ 20 years from first employment (o/e = 1.67, based on 1 case (Paci *et al.*, unpublished)).

In a cohort of 1165 men with occupational exposure to benzene in rubber hydrochloride manufacture during the period 1940–65, there was a three-fold risk of leukaemia mortality, based on nine cases, in follow-up from 1950 (or date of first employment in an exposed department, if later) to the end of 1981

(Rinsky et al., 1987). No analyses of risk according to date of death were presented, but, on examining the dates of death of the leukaemia cases, eight were during 1950–61, and only one subsequently (in 1979). Three of four other deaths due to leukaemia (on best evidence) in employees at the plants during the study years, but not within the strict cohort study definitions, were during 1950–59, and one in 1984 (Rinsky et al., 1981, 1987). Exposure levels at the plants appear to have decreased over time (Rinsky et al., 1981, 1987), but the exact pattern of decrease is unclear.

Aksoy (1985) noted that the crude incidence rate of leukaemia in shoe workers in Istanbul from 1967 to 1974 (as ascertained apparently from selected hospitals) was over twice that stated to occur in the general population. Before 1969, shoe workers had been chronically exposed to benzene, but benzene use was gradually discontinued from 1969 onwards. The numbers of leukaemic shoe workers seen annually at the hospitals increased to a peak of several cases per year in the early 1970s and then decreased so that only one case was seen during 1976–83. Although this is compatible with effective prevention, interpretation must be tentative because no analyses were presented of risks over time or of person-years at risk, no allowance was made for age, and the method of ascertainment of cases may have been incomplete.

A decrease in risk of lung cancer mortality over time in a cohort of men who were iron-ore miners in Cumbria in September 1939 may reflect several of the preventive mechanisms discussed above. Relative risk was substantially raised in 1948–67 (o/e = 1.59; obs. = 50), attributed to either radioactivity in the air underground or exposure to iron oxide, but only slightly raised during later follow-up from 1968 to 1982 (o/e = 1.13; 26 cases)[1] (Kinlen & Willows, 1988). Factors which may have contributed to a decrease were: improved dust control and ventilation at the mines; half of the man-years after 1967 were contributed by men over usual retirement age; after 1967 many men left the mines before the usual age of retirement, notably when the largest mine, with highest radon levels, closed in 1969; and, in more recent years, fewer men have been left of those who had long experience of high exposure before the ventilation improvements started in 1935 had had a substantial effect.

Decreases in relative risk with time or with duration since first employment, such as those above, can give evidence for effectiveness of prevention if a multiplicative model is the best description of the effect of occupational exposure on the underlying risk in the general population, but not necessarily under some other models. Data on risk of lung and nasal cancer mortality in a cohort of nickel refinery workers in South Wales (Table 2) illustrate this. In men first employed before 1925 at the refinery, relative risks for lung and for nasal sinus cancer mortality were greatly raised. The relative risk for lung cancer in these men decreased with increasing duration since first employment, suggesting, under a multiplicative model, a preventive effect of the apparent virtual cessation of carcinogenic exposures by 1930 (details of exposure changes are given in a later section). Nasal

---
[1] Risk was not greatly raised in 1939–1947 (o/e = 1.26), but it was based on fewer cases (8).

cancer relative risks in the cohort changed little from 20 years onwards in adjusted analyses in the table, suggesting under a multiplicative model effective prevention of an early stage carcinogen. Absolute excess risks in the cohort (Table 2), however, only began to decrease appreciably for lung cancer beyond 49 years follow-up, and for nasal cancer they increased greatly throughout follow-up; under an additive model (for which there is some evidence for the relation of nickel refinery work and smoking to lung cancer, at least (Magnus et al., 1982)), the results would give only modest evidence for prevention of lung cancer and none for prevention of nasal cancer.

Analyses for these nickel refinery workers also illustrate the effect that consideration of other time- and exposure-related variables can have on apparent decreases in risk by time since first exposure. Whereas the unadjusted relative risk of nasal cancer in the cohort, shown in Table 2, decreased by over two-fold from 20–29 years after first exposure through to ≥ 50 years after entry, relative risks adjusted for age at first employment, year of first employment, and duration of work in high risk areas (also shown in the table), scarcely changed at all with time from 20 years post-entry onwards.

A further model-dependent aspect of interpretation of time trends in risk arises for smoking-related cancers when smoking habits change over time. If smoking and an occupational risk factor both affect risk of a cancer, and if, say, the study (occupational) cohort and the comparison (general) population have smoked to a similar extent, then under a multiplicative model for the combined effects of the occupational exposure and smoking, similar changes in smoking behaviour in the study cohort and the general population will not in themselves alter relative risks over time for the occupational exposure. Under an additive model for the combined effects of smoking and the occupational factor, however, this would not be so: an equal increase in smoking by the study cohort and the general population, for instance, would itself lead to a secular reduction in relative risks for the occupational factor under study (and hence potentially give a misleading impression of successful occupational prevention).

Decreased exposure levels over time, at a workplace or for an occupational group, leading to lower disease risk in those employed recently than in those employed longer ago

Several of the best documented instances of effective prevention of occupational carcinogenesis are of this type. Since such evidence rests on comparison of different groups of individuals at different times, however, interpretation needs to be cautious – the possibility of secular changes in factors such as selection into the occupation, and confounding exposures, needs to be considered.

The classic, first example of determination of an occupational cause of cancer was that of scrotal cancer in chimney sweeps, and probably it was also the first to be prevented. Pott (1775) described the association and only three years subsequently rules were introduced in Denmark obliging sweeps to bath daily (Searle, 1976). In England and Wales preventive measures came later; it was not until 1840 that an Act was passed to prohibit sending boys up chimneys, and the practice nevertheless continued in many places for 20 or more years (Doll, 1975;

Waldron, 1983). The numbers of cases of scrotal cancer reported in sweeps after about 1880 decreased (Waldron, 1983), but whether the decrease was due to this prohibition is unclear. Risk in sweeps in England was still extremely high, and greater than in any other occupation, in 1911-38 (Henry, 1946); presumably it has much decreased since, but I have no evidence of this. More rigorous examples of reductions in risk are available, as follows, for more recently discovered occupational hazards.

Lung and nasal sinus cancers in nickel refinery workers

Risks of mortality from lung and nasal sinus cancers in men who had been employed at a nickel refinery in South Wales were extremely high for those first employed early in this century, and reduced greatly in later entrants (Table 3).[2]

The changes in risk correspond well with changes in process and hygiene at the factory: the calciners were altered in 1911 such that operators were exposed to more dust than previously (Doll et al., 1977). From 1921, arsenic exposures were rapidly reduced, in 1924[3] cotton respirator pads were introduced and rapidly accepted by a high proportion of men (Doll et al., 1977), and in 1924 the calciners were altered with reduction in dust emissions; after 1932 the amount of copper in the raw material was reduced by about 90% and the sulfur almost completely removed (Peto et al., 1984); there were also various changes to process chemistry and equipment after 1930 including the installation of new calciners between 1931 and 1936 (Kaldor et al., 1986). The major reduction in nasal cancer risk earlier than that for lung cancer may be because the cotton respirator pads introduced in 1924[3] were particularly effective for large particles which otherwise tended to deposit in the nose.

Risks of nasal and lung cancer in a Norwegian nickel refinery were also greatly increased in workers from the early years of operation (Table 3). Risk of nasal cancer was over 100-fold raised for those first employed in 1916-29, and decreased steadily in succeeding cohorts, with a similar decrease within strata by years since first employment (Magnus et al., 1982). Major changes in processes, that led to reduced exposure to dust and fumes, had occurred over the years, particularly since 1950 (Pedersen et al., 1973). For lung cancer, however, relative risks did not show a clear decrease between several of the successive cohorts, although there was a decrease when comparing entrants from 1930 onwards with those entering before that date. The expected numbers of lung cancers in the analysis were based on national rates, which increased sharply over time. Recalculation of the lung cancer relative risks with an adjustment for smoking based on an additive assumption for smoking and nickel exposure risks (the model most closely suggested by data from the study) still showed no clear indication of a decrease in risk in succeeding cohorts entering between 1930 and 1959 (Magnus et al., 1982).

---

[2] The pattern of results for the pre-1925 entrants, adjusted for age at first employment, time since first employment, and duration in high-risk areas, was similar to that of the unadjusted risks shown in the table, except that the decrease in risk of nasal cancer began to appear slightly later: for 1920-1924 rather than for 1915-19 entrants.

[3] 1922 according to Peto et al. (1984)

Table 3. Cancers of lung and nose and nasal sinuses in nickel refinery workers, in relation to year of first employment

(a) Mortality in males (Clydach, S. Wales) followed to end of 1981. Source: data in Kaldor et al. (1986)

Year of first employment	Lung cancer		Cancer of nose and nasal sinuses	
	o/e	obs.	o/e	obs.
<1910	8.10	23	275	11
1910-14	8.65	39	650	26
1915-19	3.16	13	300	9 a
1920-24	4.03	62	100	10
1925-29	2.01	11	$-b$	0^a
1930-44	1.20	11	$-c$	0

(b) Cancer incidence in males (Kristiansand, Norway) followed to end of 1979. Source: Magnus et al. (1982)

Year of first employment	Lung cancer		Cancer of nose and nasal sinuses	
	o/e	obs.	o/e	obs.
1916-29	8.3	11	122.8	7
1930-39	4.3	19	60.8	9
1940-49	3.6	17	17.6	3
1950-59	3.4	33	6.1	2
1960-65	1.2	2	$-c$	0

(c) Mortality in males (Copper Cliff, Canada) followed to end of 1978. Source: Chovil et al. (1981)

Year of first employment	Lung cancer	
	o/e^d	obs.
1948-51	11.36	36
1952-62	0.93	1

a One further nasal cancer was mentioned on a death certificate from this sub-cohort, but not given as the underlying cause of death
b 0.03 expected
c 0.05 expected
d Approximate – not from a full conventional analysis.
Statistical testing not published for these data.

In a cohort of nickel refinery workers from a Canadian sinter plant which operated from 1948 to 1963 (Chovil et al., 1981), mortality from lung cancer was about 11-fold raised in men first employed in 1948-51, but not raised in those who had started employment during 1952-62 (Table 3). No analyses of risk of nasal sinus cancer were presented, but it is clear from the number of cases occurring that risk must have been substantially raised. Whereas between seven and nine nasal cancers (the exact number is not clear) occurred up to May 1980 in the 279 men first employed from 1948–51, at most two (and at minimum zero) such cancers occurred in the 216 men first employed from 1952–62. It was stated that this difference could not be accounted for by differences in age structure or potential latency. Exhaust air samples indicated a four-fold decrease in airborne nickel sulfide levels from 1950 to 1958 (Roberts et al., 1984). Other than dust decreases, however, Chovil et al. (1981) could not identify any process changes.

At a US nickel refinery which had received nickel matte from smelters in Ontario[4] during the period 1922–47, but where, in late 1947, the calciners of the refinery were torn down (Enterline & Marsh, 1982b), workers with one or more years of refinery employment before 1948 (and therefore with probable exposure to calcining) showed a highly significant excess of sino-nasal cancer mortality in follow-up to the end of 1977 (o/e = 24.43 for respiratory cancers other than lung, pleura and larynx, based on two cases of sino-nasal cancer). Among other workers employed for at least a year at the plant before 1948 (in nickel alloy production, with or without under a year in the refinery) no deaths were certified to sino-nasal cancer, but two deaths occurred which on review were probably due to the tumour (0.56 expected from respiratory cancers other than lung, pleura and larynx). However, in workers who had worked for a year or more at the plant and who were first employed after 1946 (and not exposed to calcining operations), there were no deaths from sino-nasal cancers (0.11 expected from respiratory cancers other than lung, pleura and larynx). Lung cancer SMRs were not greatly raised in the earlier or in the later employee cohorts.

Cancers in workers exposed to ionizing radiation

Two categories of radiation-exposed workers provide further historic examples of greatly increased risks of cancer which diminished after preventive measures. Among women in the radium dial-painting industry in the United States (discussed in an earlier section), those first employed before 1925 suffered risks of bone cancer mortality over 100-fold greater than in the general population; the risks in these early cohorts were an order greater than for women first employed in the industry from 1925–29 (Table 4). There were also substantial differences between cohorts employed at these periods for mortality from leukaemia and blood diseases (and for 'other and unspecified cancers' (ICD7 198–199)). The reduction in risk coincided with changes in work regulations in the industry, notably the prohibition in 1925–26 of tipping or pointing of brushes between the lips, which resulted in greatly reduced radium doses (Polednak et al., 1978).

Early radiologists were exposed to high levels of radiation and accordingly suffered high risks of cancer. Smith & Doll (1981) found a significant 75% excess of cancer in male British radiologists who had entered the profession before 1921 (compared to all doctors), and significant excesses (compared to social class I) particularly for cancer of the skin and leukaemia, but also for cancers of the pancreas and lung (Table 4). They found no significant excess of cancer, however, in radiologists entering after 1920 (a non-significant 25% excess of cancers overall, with non-significant increases for the four sites above for 1921–35 en-

---
[4] These smelters also supplied the other refineries discussed above.

Table 4. Mortality from various cancers in workers exposed to ionizing radiation, in relation to year of first exposure

(a) Women working in US radium dial-painting industry followed to end of 1975. Source: Polednak et al. (1978)

Year of first employment	Bone cancer[a]		Leukaemia and blood disease	
	o/e	obs.	o/e	obs.
1915-19	233.33***	7	7.40*	2
1920-24	153.85***	20	3.33*	4
1925-29	10.00	1	1.03	1

(b) Male British radiologists. Followed to end of 1976. (Source: Smith & Doll, 1981). One sided p values in direction of the difference.

Year of first registration with radiological society	Cancer of pancreas		Lung cancer		Skin cancer		Leukaemia	
	o/e	obs.	o/e	obs.	o/e	obs.	o/e	obs.
<1921	3.23*	6	2.18*	8	7.79***	6	6.15**	4
1921-35	1.47	4	1.12	13	3.51	2	2.33	3
1936-54	-	0[b]	0.82	10	-	0[c]	0.77	1

(c) Male US radiologists. Followed to end of 1969. Age- and time-adjusted mortality rates per 1000 person-years[d]. Source: Matanoski et al. (1975)

Year of joining speciality society	Skin cancer			Leukaemia and aleukaemia		
	Radiologists	Physicians	Ophthalmologists & otolaryngologists[e]	Radiologists	Physicians	Ophthalmologists & otolaryngologists[e]
1920-29	0.31	0.01	0.03	0.44	0.17	0.05
1930-39 (through age 74)	0.12	0.02	0/0.04	0.44	0.12	0.09/0.13
1940-49 (through age 64)	0.05	0	0/0.06	0.05	0.06	0.08/0

* $p<0.05$; ** $p<0.01$; *** $p<0.001$

[a] Includes cases with underlying cause coded to radiation accidents and sequelae, but bone cancer mentioned and would have been underlying cause if radiation exposure had not been on the certificate

[b] 2.28 expected

[c] 0.44 expected

[d] Numbers of cases and statistical testing for this table not published

[e] For ophthalmologists and otolaryngologists combined in 1920-29 cohort, separately in later cohorts

trants, and no increase overall or for the four particular sites for 1936-54 entrants). The first recommendations of the British X-ray and Radium Protection Committee were issued in July 1921, and were probably fairly quickly effective in reducing exposure (Court Brown & Doll, 1958). Although the data were compatible with elimination of cancer risk in recent cohorts, follow-up of these men was insufficient to exclude appreciable raised risk (Smith & Doll, 1981).

The general adoption of safety measures by radiologists in the USA appears to have occurred later than in Britain (Seltser & Sartwell, 1965). Correspondingly, data from United States radiologists show a pattern fairly similar to Britain but with apparent later elimination of raised risks – several-fold raised risks of skin cancer, leukaemia and aleukaemia (and perhaps of brain and liver cancers, but not of lung cancer) were present for doctors who joined the US radiologists' specialty society in 1920-29 and 1930-39 (in comparison with other physician specialists), but not for those joining 1940-49 (Table 4) (Matanoski et al., 1975); excesses of lymphoma and myeloma were noted in the two later cohorts, however.

Lung cancer and mesothelioma in workers exposed to asbestos

Lung cancer mortality risk in long-serving workers at an asbestos textile factory in Rochdale, England, was far greater for men who had at least 10 years' service before 1933, when the first Asbestos Industry Regulations came into force, with very substantial reductions in dust levels (Peto, 1980; Peto et al., 1985), than for men who commenced employment in 1933 or later (Table 5) (Peto et al., 1985). Men with some, but less than 10 years, experience before 1933 did not overall show higher risk than post-1932 entrants (they did show higher risk in an analysis restricted to 20-34 years follow-up, but not in $\geq$ 35 years follow-up): possibly, therefore, exposure changes many years before 1933 may have been of importance. In analyses comparing those first employed in the years 1933-50 with 1951-74 entrants, there was not overall evidence of reduced lung cancer risk in the later cohort,[5] despite decreases in estimated exposure in some areas from 1933-50, and large reductions in measured particle counts in many areas from 1951-60. Potential explanations for the lack of improvement in risk included chance, and the possibility that less decrease in exposure may have occurred for the most carcinogenic fibres than was suggested by overall particle counts. An earlier follow-up of the same cohort (Peto, 1980) gave no indication that reductions in mesothelioma risk had occurred in succeeding cohorts, although numbers of mesotheliomas were small. The available evidence suggested that this was not primarily because of poorer diagnosis of mesotheliomas in earlier years; possible explanations included those noted above.

---

[5] In smaller female cohorts who had worked in scheduled areas for at least 10 years, the o/e for lung cancer mortality was 2.77 (obs. = 4) for 1933-50 entrants, while 0 deaths occurred from this cause (0.46 expected) in 1951-74 entrants (but expectations were based on national rates, which may have been somewhat inappropriate since a greater proportion of the workers are likely to have been regular smokers than in the general population).

In a cohort of men employed in manufacture of insulation materials using asbestos (mainly amosite), there was a two-fold risk of cancer of the lung and pleura (Acheson et al., 1984). Asbestos use at the plant had begun in 1946 or 1947; dust controls improved in 1964 when the beater floor was transferred to a new building, and in 1966 when it was enclosed; there had been subsequent improvements, and during 1978 use of asbestos was discontinued. In follow-up of the cohort to the end of 1980 (Table 5), the SMR for those entering exposure 1947-59 was somewhat greater than that for 1960-78 entrants; this was generally true also within strata by degree of exposure in most-exposed job, and duration of exposure. There were too few deaths in the post-1965 entrants to draw conclusions.

Men working during 1941-79 at a factory manufacturing friction materials with asbestos were studied by Newhouse et al. (1982). Those who had entered work at the plant before 1942, had a significant excess of deaths from cancer of the lung and pleura; there was not a significant risk, however, in men first employed 1942–50 or 1951–69 (Table 5). Seven of the eight mesotheliomas of the pleura included in the above analysis occurred in men first employed at the plant before 1931; the other was in a man first employed in 1957, but who had worked at an asbestos cement factory before 1931. Five other workers from the factory were known to have had mesothelioma; all had first been employed before 1941, but not all before 1931. There had been high asbestos concentrations in air before 1931 at the plant, but these had been much lower in 1932-50 (after the 1931 Asbestos Regulations), and there had been some further improvements subsequently.

In the USA, Blot and Fraumeni (1987) using case-control data found that the relative risk of lung cancer in shipyard workers (a group with potential asbestos exposure) was raised for men first employed up to 1946, but not for those who commenced employment later (Table 5). The lack of raised risk in more recent cohorts might, however, in part reflect that insufficient time had yet passed for any excess to emerge in these men.

Table 5. Cancers of the lung and pleura in asbestos workers in relation to date of first exposure

(a) Mortality among men working ≥ 20 years in 'scheduled' areas in an asbestos textile factory, Rochdale, UK. Followed to end of June 1983. Source: Peto et al. (1985)

Exposure	Lung cancer	
	o/e	obs.
Exposed ≥ 10 years before 1933	8.23	13
Exposed, but < 10 years, before 1933	1.76	7
First exposed 1933-50	1.45	13 ⎫
		⎬ 1.70 (17)
First exposed 1951-74	3.81	4 ⎭

Table 5 (contd)
(b) Mortality among male insulation material manufacture workers, Uxbridge, UK. Followed to end of 1980. Source: Acheson et al. (1984)

Year of first exposure	Cancers of lung and pleura	
	o/e	obs.
1947-59	2.68*	28
1960-78	1.77*	33

* $p < 0.05$. Not stated whether significant at a higher significance level

(c) Male friction materials manufacture workers, Britain. Mortality ≥ 10 years from first employment. Followed to end of 1979. Source: Newhouse et al. (1982)

Year of first employment	Cancers of lung and pleura	
	o/e	obs.
<1942	1.38*	42
1942-50	1.08	73
1951-69	0.87	36

* $p < 0.05$ (one-sided test in direction of the excess)

(d) Males in shipyard employment at several locations in USA. Case–control analysis; cases diagnosed in periods during the 1970s. Source: Blot & Fraumeni (1987)

Year of first employment	Lung cancer	
	Relative risk estimate	Nos. of cases/controls
Never	1.0	889/1112
<1930	1.6	37/35
1930-38	1.2	35/35
1939-46	1.5	181/157
1947-54	1.0	21/27
1955-64	0.9	14/19

Statistical testing not published for (a) and (d)

Lung cancer in workers exposed to arsenic

Arsenic exposure in smelter workers has also been associated with a raised risk of lung cancer, and again there is evidence of reduced risk in more recent cohorts with lower exposure (Table 6). Respiratory cancer mortality was 4.8 times that expected in men first employed at a copper smelter in Montana before 1925, but much reduced in later cohorts, with particular decreases between cohorts for 'heavy' and for 'medium' As_2O_3 exposure groups (Lee-Feldstein, 1983). Differences in age distribution between the cohorts were noted as a reason to be cautious in interpreting differences in o/e ratios between them. Around 1924, the As_2O_3 and SO_2 concentrations in the smelter atmosphere were 'supposedly greatly reduced' when the selective flotation process was introduced at the smelter; a general decline in atmospheric As_2O_3 concentrations at the smelter had occurred from 1943-65.

Table 6. Lung cancer in arsenic-exposed workers, in relation to year of first employment

(a) Mortality in male copper smelter employees, Montana, followed to end of September 1977. Source: Lee-Feldstein (1983)

Year of first employment	Respiratory cancer					
	All exposure categories		Heavy As_2O_3 exposure work areas[a]		Medium exposure work areas[a]	
	o/e	obs.	o/e	obs.	o/e	obs.
<1925	4.77**	130	7.02**	16	8.37**	51
1925-47	2.27**	129	4.46**	14	2.92**	32
1948-55	1.98**	43	2.91	3	2.70**	10

(b) Mortality in male pesticide manufacture and packaging workers, Baltimore, followed to 1977 (?end of August). Source: Mabuchi et al. (1979)

Year of first employment	Lung cancer	
	o/e	obs.
<1946	3.96*	10
1946-54	1.48	13
1955-74	–[b]	0

(c) Mortality in male copper smelter workers with > 15 years employment, Ooita, Japan. Followed to end of 1971. Source: Tokudome & Kuratsune (1976)

	Lung cancer					
	All exposure categories		Heavy exposure work areas[a]		Medium exposure work areas[a]	
	o/e	obs.	o/e	obs.	o/e	obs.
≥ 15 years employment before 1949	20.48**	17	25.00**	10	18.18**	4
15th year completed 1949-71	10.64**	5	5.56	1	17.65**	3

(d) Cancer incidence in male smelter workers (Rönnskär, Sweden), followed to end of 1982. Source: Sandström et al. (1989)

Year of first employment	Lung cancer	
	Overall follow-up o/e[c]	15-33 years follow-up o/e[c]
1928-29	2.28	2.47
1930-34	2.52	2.45
1935-39	2.05	2.38
1940-44	2.03	2.48
1945-49	1.52	1.94
1950-66	1.20	1.49

* $p < 0.05$; ** $p < 0.01$

[a] Relative exposure of work type, compared to other areas at the same time (fixed categorization of areas throughout the study)
[b] 2.4 expected
[c] As judged from data presented in the paper as a histogram only. Numbers of cases of lung cancer in each cohort were not presented, but in total there were 108 lung cancers. Significance levels were not presented.

Mabuchi et al. (1979) found a significantly raised risk of lung cancer in male pesticide workers exposed to inorganic arsenicals during manufacture and packaging of pesticides at a plant in Baltimore who were employed during 1946–74. Production of arsenical compounds was high before 1946, peaked around 1950, and was at a much lower level from 1952–65, whereas production of non-arsenicals tended to be greater in more recent years. It may therefore be the case, but is not clear, that the proportion of workers at the plant employed on work with arsenical compounds decreased over time. Hygiene at the arsenic acid plant was poor before 1952, when the plant was reconstructed and improved hygiene practices introduced. Air concentrations of arsenic in the insecticide building were thought to be at least twice as high in the 1950s as in 1972. SMRs for lung cancer decreased in succeeding entry cohorts (<1946, 1946–54, 1955–74) at the plant (Table 6). Risk was also much greater, however, in workers with 25 or more years' employment than in those with lesser durations, and since those with 25 or more years' employment at follow-up had necessarily been employed before 1952, the effects of duration of employment versus date of first employment could not fully be separated. Limited analyses for the two earlier cohorts stratifying by length of employment, for a subgroup of men who had worked predominantly in arsenic production, generally did not show lower SMRs for 1946–54 entrants than for earlier entrants, although the ratios were based on small numbers.

In a cohort of copper smelter workers in Japan employed before 1971 (Tokudome & Kuratsune, 1976) there was a significant twelve-fold raised risk of lung cancer mortality. Twenty-eight of the 29 lung cancer deaths were in men who had first worked at the smelter before 1949 (expected numbers, person-years at risk, or size of subcohorts by date of first employment were not stated). The only data on risk by date of employment (Table 6) showed substantially greater lung cancer risk for men working for at least 15 years before 1949 than for those employed 15 or more years with the 15th year completed during 1949–71. The difference in lung cancer risk between these two cohorts was particularly great for employees with heavy exposure to arsenic and other related compounds. A major reduction in copper production at the plant and in the amount of arsenic in the processed ores had occurred around the end of the second world war.

Data from the Swedish smelter discussed above in relation to evolution of risk over time (Sandström *et al.*, 1989) also give evidence of reduction in lung cancer risk when arsenic exposures are reduced. Various hygienic improvements had occurred at the smelter at different times since early in the 1930s, as discussed already. Lung cancer incidence, compared to all Swedish men, showed a progressive and substantial decrease in succeeding entry cohorts after 1934, and, although based on small numbers, there were appreciable decreases after 1944 in data referring only to 15–33 years after entry to exposure (Table 6).

Lung cancer in chromate workers

Several studies of chromate workers have published data on risk of lung cancer after preventive measures. Men who entered employment from 1932–54 at a chromate pigment factory in England and who had high or medium exposure for at least one year to lead and zinc chromates had a significantly raised risk (over two-fold) of mortality from lung cancer (Davies, 1984). Conditions at the factory had been so bad that until about 1945, approximately one worker in twelve was affected by lead poisoning during his employment. In 1955 there were major improvements to working conditions and enforcement of preventive measures, but, notably for the assessment of the effect of preventive efforts *per se*, the products and processes used did not alter and production was increasing. Workers entering employment at the factory from 1955 to mid-63 (i.e., essentially after the improvements) showed no raised risk of lung cancer in follow-up to 1981, although results were based on small numbers.

Alderson *et al.* (1981) found that relative risk of lung cancer mortality in workers followed to 1977 at a chromate-producing factory in Britain decreased from 3.0 (7 cases) for those employed only before major plant modification (at a date which is unclear, perhaps during 1957–61) to 2.0 (27 cases) for those employed both before and after the change, and 1.9 (2 cases) for those employed only subsequently (but, it should be noted, the results are based on small numbers for the pre-modification and post-modification risks). In data from three chromate-producing factories in Britain, including that above, the authors found a reduction in risk for men employed solely after plant modification, in a multivariate analysis which included several other variables such as duration of follow-up and duration of employment. Relative risks, numbers of cases in each category, or significance were not stated for this analysis, however.

One study intended to ascertain whether modernization was followed by decreased risk of lung cancer gave less optimistic results. A US chromium chemical production plant had been rebuilt, partly in 1950–51 and the remainder in 1960, to reduce exposure to chromium-bearing dusts (Hayes et al., 1979). Risk of lung cancer mortality was significantly threefold raised for long-term (3+ years) employees first employed before any of the plant rebuilding, and for those first employed from 1950–59 in the unrebuilt facility, but there was also a similar, and significant[6], raised risk for those first employed from 1950–59 and working solely in the new facility (there was insufficient follow-up to determine risk adequately in more recent employees, and risks were lower in short-term employees).

Nasal adenocarcinoma in woodworkers

An approximately 500 to 1000-fold increased risk of adenocarcinoma of the nasal cavity and sinuses has been found in woodworkers in High Wycombe, England (Acheson et al., 1968). Judging from the years of first employment of cases, the hazard appears to have continued at least until the second world war, but whether it continued after then is not clear; in surveys several years ago of cases in the Oxford region (Acheson et al., 1968) and elsewhere in England and Wales (Acheson et al., 1972) only one case was found in a woodworker known to have been first employed after 1939.

Acheson et al. (1982) estimated the incidence of nasal adenocarcinoma in succeeding cohorts of skilled furniture makers in Buckinghamshire. The cases were nasal adenocarcinomas incident during 1945 to 1981, recorded in the Oxford Cancer Registry and known to have worked in the Buckinghamshire industry. Populations at risk were estimated from industry sources with allowance for subsequent mortality. The peak incidence of nasal adenocarcinoma in these workers was estimated to be in men born 1900–1909 (and therefore probably entering the industry from about 1915–24); incidence then decreased, but not significantly, in the two succeeding cohorts. The exact agent responsible for the initial raised risk is unclear, and therefore whether levels of the agent decreased during the relevant period is also not known.

Leukaemia in workers exposed to benzene

Mortality data for shoe manufacture workers at a plant in Florence (Paci et al., 1989), discussed above in relation to the distribution of deaths by time since first employment, provide evidence on leukaemia (and aplastic anaemia) mortality by cohort of employment. High exposure to benzene had occurred at the plant from 1953–62, with improved ventilation installed from 1961–62, and

---

[6]Significant according to the text, but not according to a table, of the paper cited.

the product changed to less than 2% benzene from 1963, so that from 1965 benzene levels were very low. In a cohort of men employed at the factory from 1950–84 and followed to the end of 1984, there was a significant four-fold relative risk of leukaemia mortality and an over 10-fold risk for aplastic anaemia deaths. None of the 12 deaths from leukaemia (6) and aplastic anaemia (6) in the cohort, or the single such death (from aplastic anaemia) occurring in the corresponding female cohort, were in workers first employed after 1964 (Paci et al., 1989 and unpublished).

The study of benzene-exposed workers engaged in rubber hydrochloride manufacture, also referred to previously (Rinsky et al., 1981, 1987), provided some data on leukaemia mortality by date of first employment. In analyses of follow-up to mid-1975, there was a significant risk (o/e = 5.60, based on seven cases) for pre-1950 entrants, while for 1950–59 entrants there was an o/e of 2.17, based on one case starting work in 1950 (Rinsky et al., 1981). Of the five other known leukaemia deaths in workers at the plants, three were in pre-1950 entrants and the others in entrants in 1950 and 1952 (Rinsky et al., 1981, 1987). Exposure levels at the plants had apparently decreased over time, but the exact pattern of this decrease is not clear.

Bladder cancer in dye workers

The high risk of bladder tumours in men engaged in work with various aromatic amines in the dye industry was shown in classic studies by Case and colleagues over 30 years ago (Case et al., 1954; Case & Pearson, 1954). In follow-up to the end of January 1952, bladder papilloma or cancer had been incident in around 18% of men in cohorts entering before 1930, while more recent cohorts showed steady decreases in the proportion of men in whom bladder tumours had occurred, as would be expected from the shorter follow-up periods available. Relative risks by entry cohort were not presented. In comparisons where expectations were based on induction periods (which now seems, in retrospect, an unsatisfactory method), there was for workers exposed to benzidine, and for those with 'mixed exposures' (i.e. two or more of β-naphthylamine, α-naphthylamine and benzidine) a lower observed/expected ratio for bladder tumours among men entering from 1930 onwards than there was among earlier entrants. Alterations of plant and techniques which might have altered risk had first been made about 1935, and developed progressively afterwards, especially after 1945.

These original cohorts do not appear to have been followed up further, nor do there appear to have been any subsequent published cohort studies of dye workers in England and Wales. However, one of the major chemical companies with workers in the original cohort studied by Case has informed me that in 1981 they identified a cohort of 1285 dye workers who worked with the company during 1910–47 (probably most of the dye workers exposed to the agents studied by Case who were employed at the company during that period). By the end of 1980, 830 deaths had occurred, for which the cause could be identified from death certificates for 717 (86%). Based on the 717 deaths in this longer follow-up, 18% of the workers first employed during 1917 to May 1922 are known to have died of bladder cancer, with greatly decreasing proportions of workers in subsequent entry cohorts to 1947 known to have died of this cause (e.g. 14% of the

June 1927 to May 1932 entrants, 6% of the June 1937 to May 1942 entrants, and 2% of the June 1942 to December 1947 cohort). Although interpretation of these data must be tentative because of the unclear method by which the cohorts were recruited, the substantial proportion of deaths for which cause was unknown, the uncertain losses to follow-up, different lengths of follow-up, and lack of a person-years analysis, there is a strong suggestion, which needs more rigorous study, that a large decrease in risk of bladder cancer occurred well before epidemiological demonstration of the carcinogenic risks to dye workers, and the cessation of manufacture and handling of β-naphthylamine and benzidine by the company. The extent and timing of preventive measures and decreases in chemical exposures at the company before 1948 are not known at present (although there is general information about the industry given by Case and noted above).

Meigs et al. (1986), investigating bladder cancer incidence in men working at a benzidine manufacturing facility in Connecticut, found much higher risk for those first employed in the early years (1945–49) of the plant (o/e = 9.76; obs. = 4) than for those first employed in 1950–54 (o/e = 2.13; obs. = 1), after an equal period of follow-up (truncated after 24 years) but based on small numbers of cases. Extensive preventive measures had been instituted around 1950, with a prompt reduction to about a third of previous levels in workers' exposure to diamines.

A less satisfactory analysis suggests effective prevention of bladder tumours at another US plant. Ferber et al. (1976) found 36 benign or malignant bladder tumours incident from 1930 to May 1975 in workers at a US dye–manufacturing facility exposed mainly or solely to benzidine with first exposures before 1955, and 79 in other workers at the plant first exposed before 1955. In that year, a new fully enclosed benzidine process was installed and hygiene controls were introduced; major process improvements in other areas of the facility were also made at about the same time. In workers first exposed from 1955 onwards, no bladder tumours had been reported to May 1975 (incidence appears to have been ascertained from medical reports and compensation claims, but it is unclear how complete the bladder tumour ascertainment and the follow-up were). No person-years at risk data, or expected numbers of bladder tumours, or observed/expected analyses based on person-years, were presented.

Bladder cancer in rubber workers

Further examples of effective prevention come from the rubber industry. Among 40 867 men employed in the British rubber industry (including cable-making) and studied by Baxter and Werner (1980), an excess of bladder cancer mortality was found in men who had entered work during or before 1949 in factories where exposure to antioxidants based on 1- and 2-naphthylamine may have occurred, but not in men who had joined the industry after that year, when the compounds were withdrawn, or in men who had worked in factories where these compounds had never been used (Table 7). The excess resulted from bladder cancers in the non-tyre sector (Table 7). Based on smaller numbers, there were such excesses (i.e. in the pre-1950 potentially-exposed group only) in some tyre factories and some non-tyre sectors and occupations, including a significant risk (o/e = 3.97) for men in the extruding, calendering and other such areas, who would have been exposed to fumes from the suspect antioxidants.

Another large study of British rubber workers (which included 18% of the men in the above study) found some indication of raised risk for some groups of workers (those most likely to have been exposed to fumes from heated rubber compound) entering before 1951, but no substantial evidence of raised risk for entrants from 1951 onwards (Table 7) (Parkes et al., 1982).

Table 7. Bladder cancer mortality in rubber workers, in relation to year of first employment

(a) Male British rubber workers followed from 1 February 1967 to end of 1976. Source: Baxter & Werner (1980)

Year of first employment	Bladder cancer			
	Total industry		Non-tyre sector	
	o/e	obs.	o/e	obs.
<1950 in a factory which had used suspect antioxidants	1.44*	36	1.98**	23
≥1950 in a factory which had used suspect antioxidants	1.08	24	1.12	15
Factories which had never used suspect antioxidants	0.97	13	0.84	11

(b) Male British tyre manufacturing workers: sub-groups most likely to have been exposed to fumes from heated rubber compound. Followed to end of 1975. Source: Parkes et al. (1982) (some overlap with Baxter & Werner cohort – see text)

Year of first employment	Bladder cancer	
	o/e	obs.
1946-50	2.00	10
1951-55	_a	0 } 0.83 (2)
1956-60	4.00	2

(c) Male rubber workers, Akron, Ohio. Followed to various dates, end of 1971 to end of June 1974, depending on sub-category. Source: Monson & Nakano (1976)

Year of first employment	Bladder cancer			
	All employees		Employees who worked ≥ 25 years	
	o/e	obs.	o/e	obs.
<1925	1.40	24	1.40	21
1925-34	1.40	19	1.20	9
≥1935	0.60	5	0.60	1

* $p < 0.05$; ** $p < 0.005$
a 1.9 expected
Statistical testing not published for (b) and (c)

In a cohort of US rubber workers (Table 7) (Monson & Nakano, 1976), risk of bladder cancer mortality was raised for those first employed before 1935, but not for workers first employed later (although the results were based on insufficient person-years in the latter category to be confident that the risk had been eliminated). Use of products containing β-naphthylamine may not have been discontinued at the plant until the late 1940s (the exact date is not clear).

For more recently discovered occupational carcinogens, there is as yet much shorter follow-up since the time when consequent preventive measures were undertaken, and therefore there is very limited evidence about the effectiveness of these measures:

Angiosarcoma of the liver in workers exposed to vinyl chloride

The occupational risk of hepatic angiosarcoma in workers exposed to vinyl chloride monomer is an example where preventive measures were instituted promptly and stringently after the hazard was identified. The first report of carcinogenesis was published in 1974 by Creech & Johnson; there was very rapidly a several-fold reduction in exposure levels in the UK, with further reductions subsequently (Jones *et al.*, 1988b). The minimum induction period observed for such tumours is now beginning to be exceeded by the period of follow-up available; so far no cases of angiosarcoma of the liver have been observed in the UK in workers first exposed after 1974 (Jones, R.D., personal communication), and none in workers first exposed after 1969 had been notified to a worldwide registry by the end of 1984 (Forman *et al.*, 1985).

Lung cancer in workers exposed to chloromethyl methyl ether

Raised risk of lung cancer in men exposed to chloromethyl methyl ether (CMME) has also been identified only recently (Figueroa *et al.*, 1973), with limited follow-up of risk in workers for whom preventive measures were then undertaken. McCallum *et al.* (1983), following until the end of 1980 workers at a UK factory which had manufactured CMME, found a significant excess of lung cancer in the workers studied overall. The process had been altered in 1971–72 with great reduction in CMME exposure, and the authors noted no deaths from lung cancer (yet) in workers entering employment after March 1972. No raised risk of lung cancer was found in another CMME factory where it was thought CMME levels had always been low (McCallum *et al.*, 1983). Others, in the USA, have similarly found raised risk restricted to a firm with high chloromethyl ether exposure levels (Pasternack *et al.*, 1977). The development of respiratory cancer mortality risk over time in a cohort first employed with chloromethyl ethers before 1972 (Maher & DeFonso, 1987) has already been discussed above.

Less well established etiology

When occupational etiology of a cancer is not clearly established, a decline in risk in more recent employment cohorts is initially most interesting as evidence of causation. Thus, the SMR for lung cancer in a factory manufacturing antimony oxide was over twice as great for men first employed before 1961 (excluding men who ceased employment before then, for whom records were not available) than for men first employed later (Newell, unpublished, cited by Doll, 1985). Dust levels had been greatly reduced since the 1960s (but no dust measurements were available before 1961). Such a reduction in risk is at least favourable to a hypothesis of etiology, although not conclusive.

Decline in an occupational cancer in routine statistics

Potentially, routine statistics can be used to monitor the overall effect nationally of preventive measures, and they might give the first evidence of effects of recent measures concerning common occupational carcinogens, for which routine data might provide larger numbers than ad hoc studies. Three sources of statistics might be of value. Firstly, secular trend data on incidence or mortality from cancers which are often of occupational origin (although there are few such tumours). Thus the decline in scrotal cancer mortality in England and Wales over the last 55 years (Figure 1), and the decline in registrations of scrotal cancer for the shorter period for which data are available, presumably reflects a decreasing rate of incidence of cases of occupational etiology in the country, since most cases appear to be of occupational origin (Waldron et al., 1984). Investigation based on routine data can sometimes take the evidence further: Waldron et al. (1984) investigated secular trends by occupation in West Midlands registrations of cancer of the scrotum and found a secular decrease in cases due to pitch and tar, largely resulting from the preventive effect of the closure of a large tar distillery in the region.

Figure 1. Age-standardized mortality rates[a] for cancer of the scrotum in males, England and Wales 1911–20, 1923–49, 1958–85[b]

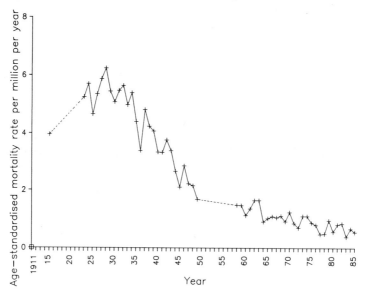

[a] Direct age-standardization, using England and Wales population 1981 as the standard.
[b] Annual data for cancer of the scrotum separated from other diseases are not available before 1923 or from 1950–57. First data point is the mean for 1911–20.

Routine secular trend statistics can also monitor prevention of mesothelioma, and perhaps of hepatic angiosarcoma (although in the UK at least, cases due to known occupational causes are a minority (Forman et al., 1985)); bladder cancer, adenocarcinoma of the nasal sinuses and non-melanoma skin cancers might in some circumstances also be possibilities for such an approach. Some evidence from this source of effective prevention of asbestos-related cancers may now be emerging: annual mesothelioma mortality rates in women in Great Britain have recently ceased to rise – perhaps a prelude to a decrease (Jones et al., 1988a); this may reflect improvements in the manufacturing sector where most women in asbestos work were employed, and also cessation of asbestos exposure for women in gas mask manufacture and other wartime work.

A second type of routine statistic which could have use for monitoring prevention is occupational mortality and cancer registration data. The usual provisos on the extent of comparability of these data at different periods in time need to be observed (e.g., differences in coding, in comparison groups, in age groups included, etc., and where SMRs are used, the strict non-comparability of these when based on different populations), but nevertheless there may be some limited value in the approach. Table 8 shows SMRs over time for lung cancer in gas workers and in doctors, and nasal cancer in woodworkers. The effect of reduced smoking by doctors compared to other men over time can clearly be seen[7]; also, perhaps, a reduction in risk for coal gas and coke makers, possibly related to

Table 8. Secular trends in standardized mortality ratios for lung cancer in male gas workers and medical practitioners, and nasal cancer in male woodworkers, England and Wales[a]

Cancer site, and occupation[b]	1930-32	1949-53	1959-63	1970-72	1979-80, 1982-83
	age 20-64 SMR (no.)	age 20-64 SMR (no.)	age 15-64 SMR (no.)	age 15-64 SMR (no.)	age 20-64 SMR (no)
Cancer of the lung (plus pleura in the three earliest periods) in makers of coal gas and coke	367 (11)	129 (63)	152 (97)	178 (36)	-
Cancer of the lung (plus pleura in the three earliest periods) in medical practitioners	137 (11)	50 (46)	48 (78)	32 (31)	22 (27)
Cancer of the nose and nasal sinuses in woodworkers	-	-	200 (12)	225 (10)	163 (6)[c]

[a] Source: Registrar General/OPCS (1938, 1958, 1971, 1978 and unpublished)
[b] Exact site and occupation categories vary a little over the years, according to changes in the ICD, in OPCS occupation coding, and in the tabulations available
[c] Recalculation from unpublished OPCS data

---

[7] It is of interest that the decrease in SMRs appears to have begun earlier than could be attributed to the publication of epidemiological evidence of the relation of smoking to lung cancer.

reduced occupational exposures. The data do not give convincing evidence nationally, however, of reduced risk of nasal cancer in woodworkers (cohort data for Buckinghamshire, where the risk was much greater than in any of these national figures, have already been discussed).

Finally, statistics on notifications of occupational cancer, compensation payments and death benefits may also give some intimation from routine sources of the effectiveness of prevention. Deaths from primary urinary tract epithelial neoplasms in workers making or handling chemicals such as β-naphthylamine and auramine, and for which industrial death benefit was paid in Great Britain, increased from 5.4 per annum in 1953–59, to 16.0 per annum in 1965–69, but have since decreased fairly steadily to 10.6 per year in 1980–86.[8] Such benefits are paid when an individual who dies from a prescribed cancer has been involved in an occupation deemed to be potentially etiological, and the worker's dependents have made a claim which has been accepted. Decreases in benefit cases, in the absence of changes that make claims more difficult, could give supplementary information of relevance to the effects of prevention, but there appears to be no experience of whether such data are in fact of value in this context.

Selective employment of low risk individuals, and preventive effects of non-occupational measures for synergistic risk factors

As well as the more obvious prevention of occupational cancers by ceasing manufacture or ceasing use of a substance, or by reducing exposures, there is also potential to reduce risk by selective employment policies which ensure that persons who work on a particular process are low-risk individuals for the tumour involved, or by altering non-occupational risk factors for a tumour which is also caused by occupation. With regard to the former, if risk of a cancer is particularly high, say, in persons who start a particular employment when young (as has been found for bladder cancer in one study (Hoover & Cole, 1973) but not in another (Monson & Nakano, 1976)), then selective employment of older individuals to work on these processes could reduce risk. Selective employment of older persons would presumably also reduce overall morbidity for tumours of long latency. Although such policies have sometimes been pursued – employment of young persons in arsenic-handling workplaces was at one time prohibited in several countries (Neubauer, 1947), and older men were selectively employed in some dyestuffs work (Gadian, 1975) – I can find no evidence of the preventive effectiveness of such measures.

Reduction in levels of risk factors synergistic with occupational factors – for instance smoking as a multiplicative risk factor with asbestos (Saracci, 1977) and with radon[9] for lung cancer – should in theory be a powerful method to reduce risk for occupational exposures. The only substantial evidence I can find on the effectiveness of such preventive measures is that, in long-term asbestos insulation

---

[8] Data from unpublished figures kindly supplied by SR8 branch, Department of Social Security.

[9] In one analysis (Archer et al., 1976), although not in another (Radford & St Clair Renard, 1984).

workers followed for 10 years, far lower relative risks of lung cancer mortality (less than half) occurred in men who at the start of the follow-up period had been ex-smokers for at least five years than in those who were current smokers at the start of follow-up with the same level of cigarette use (Hammond et al., 1979). Risk in the ex-smokers, however, was not reduced to the level in non-smokers. (Men who had ceased smoking less than five years before the start of follow-up did not have impressive reductions in risk, presumably because some of those ill with lung cancer selectively stopped smoking. For about a third of the cohort, the investigators had no data on smoking).

Decrease in incidence of cancer precursors

For many occupational carcinogens, the induction period from first exposure to incidence of cancer is typically very long, and it therefore takes many years to obtain evidence of effectiveness of prevention. In such circumstances, initial evidence of biological effectiveness of the measure and of potential reduction in cancer risk might come from monitoring incidence of precursors of the cancer. Thus, incidence of abnormalities detected by urinary cytological assays (which, since 1967, have by regulation been offered six-monthly to workers in the UK exposed to carcinogenic aromatic amines) or of chromosome aberrations and sister chromatid exchange in peripheral lymphocytes might show an effect of prevention before evidence could be found of decreases in mortality from the corresponding cancers. Again, however, I know of no examples where such evidence has suggested that preventive measures have succeeded.

Conclusions

It is striking that most of the clearest examples of prevention of occupational cancer relate to decreases from very large risks of cancer incurred many years ago – for instance, radiologists before 1921, nickel workers before 1925, radium dial-painters before 1925. Partly, of course, this reflects the long induction period between first exposure and cancer incidence, which is characteristic of many of the known occupational carcinogens. Therefore in some instances, such as risk of hepatic angiosarcoma in those exposed to vinyl chloride monomer, although there is not yet clear evidence that prevention has been effective, one can be reasonably certain that such evidence will become plain in time. It should be noted, however, that when occupational cancers occur at uncommon sites and in unusual occupational groups, it may be very difficult, and it may take many years, even for high relative risks, to obtain clear results about whether preventive measures have been effective. For example, the evidence of raised risk of cancer of the nasal sinuses in workers engaged in manufacture of isopropyl alcohol by the strong-acid process appears to be based on only seven cases reported worldwide in the literature in over 30 years (Weil et al., 1952; Eckhardt, 1974; Alderson & Rattan, 1980); evidence of successful prevention of this risk may take many years to accumulate.

The lack of clear recent examples of effectiveness of prevention also reflects, however, the much greater difficulty of detecting decreasing risk in circumstances where relative risk is not massive: greater sophistication of both data collection and analysis are needed in such instances:

Firstly, there is a need for quantitative data to be collected on dose and duration of exposure, if possible covering the whole of an individual's relevant occupational career. Ideally, this should not just be for environmental levels of the carcinogen, but would also include levels in biological samples, and preferably in the tissue concerned; for instance, although urinary concentrations of nickel may give a good indication of recent nickel exposure (Tola et al., 1979), there is a poor correlation between urinary nickel concentrations and those in the nasal mucosa in active workers (Torjussen & Andersen, 1979). Measurement of exposure levels in relation to prevention is further complicated in many instances (e.g., woodworking and nasal cancer) by the fact that the specific carcinogen has not been identified and therefore it is unclear just what should be measured.

Secondly, there is a need to collect information on confounding variables from occupational cohorts: in particular, to collect information on smoking. Lung cancer represents the majority of known occupationally induced cancer deaths (Doll & Peto, 1981), and if modest decreases in lung cancer in relation to prevention are to be detected, the need to be able to account for changes in smoking behaviour in a cohort will become critical.

Finally, in analyses there is the need for statistical testing of changes in risk after preventive measures[10], and for greater sophistication in methods to disentangle the simultaneous, often highly interrelated, effects of time- and exposure-related variables such as year of entry to an occupation, age at entry, time since entry, duration for which the occupation has been followed, and calendar period (see Thomas, 1987), and to allow for induction periods. Clearly, for instance, those who were first exposed the longest ago in calendar time to historic high levels of a carcinogen are also those for whom greatest follow-up is available, and therefore, if a long induction period is required for the appearance of a cancer, the lack of a raised risk in more recent cohorts may indicate merely that these cohorts have not yet passed the induction period. The effect that allowance for time- and exposure-related variables can have is demonstrated by the analyses of mortality in nickel refinery workers (Peto et al., 1984; Kaldor et al., 1986) already discussed.

In analyses which have dealt with major decreases in risk many years ago, the aspects above may be relatively unimportant; however, as interest focuses increasingly on more modest effects, attention to such issues will become critical.

Acknowledgements

I am grateful for helpful discussions to Dr V. Beral, Dr A.J. Fox, Professor M. Harrington, Mr J. Hodgson, Dr R. Jones, Dr L.J. Kinlen, Dr G. Parkes, Professor J. Peto and Dr J.A.H. Waterhouse. I also thank Mr S. Gillom for data on death benefits.

---

[10] For some of the very major reductions in risk cited in this paper such formal testing seems unnecessary, but for some of the less substantially-based decreases cited, chance does not appear to be a very unlikely explanation; in particular, a decrease from a significant to a non-significant raised risk does not of itself constitute a significant reduction.

References

Acheson, E.D., Cowdell, R.H., Hadfield, E. & Macbeth, R.G. (1968) Nasal cancer in woodworkers in the furniture industry. *Br. Med. J.*, *ii*, 587-596

Acheson, E.D., Cowdell, R.H. & Rang, E. (1972) Adenocarcinoma of the nasal cavity and sinuses in England and Wales. *Br. J. Ind. Med.*, *29*, 21-30

Acheson, E.D., Gardner, M.J., Winter, P.D. & Bennett, C. (1984) Cancer in a factory using amosite asbestos. *Int. J. Epidemiol.*, *13*, 3-10

Acheson, E.D., Winter, P.D., Hadfield, E. & Macbeth, R.G. (1982) Is nasal adenocarcinoma in the Buckinghamshire furniture industry declining? *Nature*, *299*, 263-265

Aksoy, M. (1985) Malignancies due to occupational exposure to benzene. *Am. J. Ind. Med.*, *7*, 395-402

Alderson, M.R. & Rattan, N.S. (1980) Mortality of workers on an isopropyl alcohol plant and two MEK dewaxing plants. *Br. J. Ind. Med.*, *37*, 85-89

Alderson, M.R., Rattan, N.S. & Bidstrup, L. (1981) Health of workmen in the chromate-producing industry in Britain. *Br. J. Ind. Med.*, *38*, 117-124

Archer, V.E., Gillam, D. & Wagoner, J.K. (1976) Respiratory disease mortality among uranium miners. *Ann. NY Acad. Sci.*, *271*, 280-293

Baxter, P.J. & Werner, J.B. (1980) *Mortality in the British rubber industries 1967-76*. London, Her Majesty's Stationery Office

Beral, V., Fraser, P., Booth, M. & Carpenter, L. (1987) Epidemiological studies of workers in the nuclear industry. In: Russell Jones, R. & Southwood, R., eds, *Radiation and Health: the Biological Effects of Low-level Exposure to Ionizing Radiation*, Chichester, John Wiley, pp. 97-106

Blot, W.J. & Fraumeni, J.F., Jr (1987) Time-related factors in cancer epidemiology. *J. Chron. Dis.*, *40*, Suppl. 2, 1S-8S

Bond, G.G., Shellenberger, R.J., Fishbeck, W.A., Cartmill, J.B., Lasich, B.J., Wymer, K.T. & Cook, R.R. (1985) Mortality among a large cohort of chemical manufacturing employees. *J. Natl Cancer Inst.*, *75*, 859-869

Case, R.A.M., Hosker, M.E., McDonald, D.B. & Pearson, J.T. (1954) Tumours of the urinary bladder in workmen engaged in the manufacture and use of certain dyestuff intermediates in the British chemical industry. Part I. The role of aniline, benzidine, alpha-naphthylamine and beta-naphthylamine. *Br. J. Ind. Med.*, *11*, 75-104

Case, R.A.M. & Pearson, J.T. (1954) Tumours of the urinary bladder in workmen engaged in the manufacture and use of certain dyestuff intermediates in the British chemical industry. Part II. Further consideration of the role of aniline and of the manufacture of auramine and magenta (fuchsine) as possible causative agents. *Br. J. Ind. Med.*, *11*, 213-216

Chovil, A. (1981) The epidemiology of primary lung cancer in uranium miners in Ontario. *J. Occup. Med.*, *23*, 417-421

Chovil, A., Sutherland, R.B. & Halliday, M. (1981) Respiratory cancer in a cohort of nickel sinter plant workers. *Br. J. Ind. Med.*, *38*, 327-333

Court Brown, W.M. & Doll, R. (1958) Expectation of life and mortality from cancer among British radiologists. *Br. Med. J.*, *ii*, 181-187

Creech, J.L., Jr & Johnson, M.N. (1974) Angiosarcoma of liver in the manufacture of polyvinyl chloride. *J. Occup. Med.*, *16*, 150-151

Davies, J.M. (1984) Lung cancer mortality among workers making lead chromate and zinc chromate pigments at three English factories. *Br. J. Ind. Med.*, *41*, 158-169

Delzell, E. & Monson, R.R. (1981) Mortality among rubber workers. IV. General mortality patterns. *J. Occup. Med.*, *23*, 850-856

Doll, R. (1975) Pott and the prospects for prevention. *Br. J. Cancer*, *32*, 263-272

Doll, R. (1985) Occupational cancer: a hazard for epidemiologists. *Int. J. Epidemiol.*, *14*, 22-31

Doll, R., Mathews, J.D. & Morgan, L.G. (1977) Cancers of the lung and nasal sinuses in nickel workers: a reassessment of the period of risk. *Br. J. Ind. Med.*, *34*, 102-105

Doll, R. & Peto, R. (1976) Mortality in relation to smoking: 20 years' observations on male British doctors. *Br. Med. J.*, *2*, 1525-1536

Doll, R. & Peto, R. (1981) *The Causes of Cancer. Quantitative Estimates of Avoidable Risks of Cancer in the United States Today*, Oxford, Oxford University Press

Doll, R., Vessey, M.P., Beasley, R.W.R. et al. (1972) Mortality of gasworkers - final report of a prospective study. *Br. J. Ind. Med.*, *29*, 394-406

Easton, D.F., Peto, J. & Doll, R. (1988) Cancers of the respiratory tract in mustard gas workers. *Br. J. Ind. Med.*, *45*, 652-659

Eckhardt, R.E. (1974) Annals of industry - noncasualties of the work place. *J. Occup. Med.*, *16*, 472-477

Enterline, P.E. (1974) Respiratory cancer among chromate workers. *J. Occup. Med.*, *16*, 523-526

Enterline, P.E., Hartley, J. & Henderson, V. (1987) Asbestos and cancer: a cohort followed up to death. *Br. J. Ind. Med.*, *44*, 396-401

Enterline, P.E. & Marsh, G.M. (1982a) Cancer among workers exposed to arsenic and other substances in a copper smelter. *Am. J. Epidemiol.*, *116*, 895-911

Enterline, P.E. & Marsh, G.M. (1982b) Mortality among workers in a nickel refinery and alloy manufacturing plant in West Virginia. *J. Natl Cancer Inst.*, *68*, 925-933

Ferber, K.H., Hill, W.J. & Cobb, D.A. (1976) An assessment of the effect of improved working conditions on bladder tumor incidence in a benzidine manufacturing facility. *Am. Ind. Hyg. Assoc. J.*, *37*, 61-68

Figueroa, W.G., Raszkowski, R. & Weiss, W. (1973) Lung cancer in chloromethyl methyl ether workers. *New Engl. J. Med.*, *288*, 1096-1097

Forman, D., Bennett, B., Stafford, J. & Doll, R. (1985) Exposure to vinyl chloride and angiosarcoma of the liver: a report of the register of cases. *Br. J. Ind. Med.*, *42*, 750-753

Fox, A.J. & Collier, P.F. (1976) Low mortality rates in industrial cohort studies due to selection for work and survival in the industry. *Br. J. Prev. Soc. Med.*, *30*, 225-230

Fraumeni, J.F., Jr, Lloyd, J.W., Smith, E.M. & Wagoner, J.K. (1969) Cancer mortality among nuns: role of marital status in etiology of neoplastic disease in women. *J. Natl Cancer Inst.*, *42*, 455-468

Gadian, T. (1975) Carcinogens in industry, with special reference to dichlorobenzidine. *Chem. Ind.*, *4*, 821-831

Gubéran, E. & Usel, M. (1987) Unusual mortality pattern among short term workers in the perfumery industry in Geneva. *Br. J. Ind. Med.*, *44*, 595-601

Gustavsson, P., Gustavsson, A. & Hogstedt, C. (1987) Excess mortality among Swedish chimney sweeps. *Br. J. Ind. Med.*, *44*, 738-743

Gustavsson, P., Gustavsson, A. & Hogstedt, C. (1988) Excess of cancer in Swedish chimney sweeps. *Br. J. Ind. Med.*, *45*, 777-781

Hammond, E.C., Selikoff, I.J. & Seidman, H. (1979) Asbestos exposure, cigarette smoking and death rates. *Ann. NY Acad. Sci.*, *330*, 473-490

Hayes, R.B., Lilienfeld, A.M. & Snell, L.M. (1979) Mortality in chromium chemical production workers: a prospective study. *Int. J. Epidemiol.*, *8*, 365-374

Henry, S.A. (1946) *Cancer of the Scrotum in Relation to Occupation*, London, Humphrey Milford, Oxford University Press

Hoover, R. & Cole, P. (1973) Temporal aspects of occupational bladder carcinogenesis. *New Engl. J. Med.*, *288*, 1040-1043

Howe, G.R., Nair, R.C., Newcombe, H.B., Miller, A.B. & Abbatt, J.D. (1986) Lung cancer mortality (1950-80) in relation to radon daughter exposure in a cohort of workers at the Eldorado Beaverlodge uranium mine. *J. Natl Cancer Inst.*, *77*, 357-362

Howe, G.R., Nair, R.C., Newcombe, H.B., Miller, A.B., Burch, J.D. & Abbatt, J.D. (1987) Lung cancer mortality (1950-80) in relation to radon daughter exposure in a cohort of workers at the Eldorado Port Radium uranium mine: possible modification of risk by exposure rate. *J. Natl Cancer Inst.*, *79*, 1255-1260

Jones, R.D., Smith, D.M. & Thomas, P.G. (1988a) Mesothelioma in Great Britain in 1968-1983. *Scand. J. Work Environ. Health*, *14*, 145-152

Jones, R.D., Smith, D.M. & Thomas, P.G. (1988b) A mortality study of vinyl chloride monomer workers employed in the United Kingdom in 1940-1974. *Scand. J. Work Environ. Health*, *14*, 153-160

Kaldor, J., Peto, J., Easton, D., Doll, R., Hermon, C. & Morgan, L. (1986) Models for respiratory cancer in nickel refinery workers. *J. Natl Cancer Inst.*, *77*, 841-848

Kinlen, L.J. (1982) Meat and fat consumption and cancer mortality: a study of strict religious orders in Britain. *Lancet*, *i*, 946-949

Kinlen, L.J. & Willows, A.N. (1988) Decline in the lung cancer hazard: a prospective study of the mortality of iron ore miners in Cumbria. *Br. J. Ind. Med.*, 45, 219-224

Lancet (1965) Bladder tumours in industry. *Lancet*, ii, 1173

Lee-Feldstein, A. (1983) Arsenic and respiratory cancer in humans: follow-up of copper smelter employees in Montana. *J. Natl Cancer Inst.*, 70, 601-609

Lorenz, E. (1944) Radioactivity and lung cancer; a critical review of lung cancer in the miners of Schneeberg and Joachimsthal. *J. Natl Cancer Inst.*, 5, 1-15

McCallum, R.I., Woolley, V. & Petrie, A. (1983) Lung cancer associated with chloromethyl methyl ether manufacture: an investigation at two factories in the United Kingdom. *Br. J. Ind. Med.*, 40, 384-389

Mabuchi, K., Lilienfeld, A.M. & Snell, L.M. (1979) Lung cancer among pesticide workers exposed to inorganic arsenicals. *Arch. Environ. Health*, 34, 312-320

Magnus, K., Andersen, A. & Høgetveit, A.C. (1982) Cancer of respiratory organs among workers at a nickel refinery in Norway. Second report. *Int. J. Cancer*, 30, 681-685

Maher, K.V. & DeFonso, L.R. (1987) Respiratory cancer among chloromethyl ether workers. *J. Natl Cancer Inst.*, 78, 839-843

Matanoski, G.M., Seltser, R., Sartwell, P.E., Diamond, E.L. & Elliot, E.A. (1975) The current mortality rates of radiologists and other physician specialists: specific causes of death. *Am. J. Epidemiol.*, 101, 199-210

Meigs, J.W., Marrett, L.D., Ulrich, F.U. & Flannery, J.T. (1986) Bladder tumor incidence among workers exposed to benzidine: a thirty-year follow-up. *J. Natl Cancer Inst.*, 76, 1-8

Melick, W.F., Naryka, J.J. & Kelly, R.E. (1971) Bladder cancer due to exposure to para-aminobiphenyl: a 17-year followup. *J. Urol.*, 106, 220-226

Miller, R.W. (1988) Rare events as clues to cancer etiology: the eighteenth annual symposium of the Princess Takamatsu cancer research fund. *Cancer Res.*, 48, 3544-3548

Monson, R.R. & Nakano, K.K. (1976) Mortality among rubber workers. I. White male union employees in Akron, Ohio. *Am. J. Epidemiol.*, 103, 284-296

Neubauer, O. (1947) Arsenical cancer: a review. *Br. J. Cancer*, 1, 192-251

Newhouse, M.L., Berry, G. & Skidmore, J.W. (1982) A mortality study of workers manufacturing friction materials with chrysotile asbestos. *Ann. Occup. Hyg.*, 26, 899-909

Paci, E., Buiatti, E., Seniori Costantini, A., Miligi, L., Pucci, N., Scarpelli, A., Petrioli, G., Simonato, L., Winkelmann, R. & Kaldor, J.M. (1989) Aplastic anaemia, leukaemia and other cancer mortality in a cohort of shoe workers exposed to benzene. *Scand. J. Work Environ. Health*, 15, 313-318

Parkes, H.G., Veys, C.A., Waterhouse, J.A.H. & Peters, A. (1982) Cancer mortality in the British rubber industry. *Br. J. Ind. Med.*, 39, 209-220

Pasternack, B.S., Shore, R.E. & Albert, R.E. (1977) Occupational exposure to chloromethyl ethers: a retrospective cohort mortality study (1948-1972). *J. Occup. Med.*, 19, 741-746

Pedersen, E., Høgetveit, A.C. & Andersen, A. (1973) Cancer of respiratory organs among workers at a nickel refinery in Norway. *Int. J. Cancer*, 12, 32-41

Peto, J. (1980) The incidence of pleural mesothelioma in chrysotile asbestos textile workers. In: Wagner, J.C., ed., *Biological Effects of Mineral Fibres*, Vol. 2, (IARC Scientific Publications No. 30), Lyon, International Agency for Research on Cancer, pp. 703-711

Peto, J., Cuckle, H., Doll, R., Hermon, C. & Morgan, L.G. (1984) Respiratory cancer mortality of Welsh nickel refinery workers. In: Sunderman, F.W., Jr, ed., *Nickel in the Human Environment* (IARC Scientific Publications No. 53), Lyon, International Agency for Research on Cancer, pp. 37-46

Peto, J., Doll, R., Hermon, C., Binns, W., Clayton, R. & Goffe, T. (1985) Relationship of mortality to measures of environmental asbestos pollution in an asbestos textile factory. *Ann. Occup. Hyg.*, 29, 305-355

Polednak, A.P., Stehney, A.F. & Rowland, R.E. (1978) Mortality among women first employed before 1930 in the US radium dial-painting industry. A group ascertained from employment lists. *Am. J. Epidemiol.*, 107, 179-195

Pott, P. (1775) *Chirurgical Observations Relative to the Cataract, the Polypus of the Nose, the Cancer of the Scrotum, the Different Kinds of Ruptures and the Mortification of the Toes and Feet*, London, Hawes, Clarke & Collins

Radford, E.P. & St Clair Renard, K.G. (1984) Lung cancer in Swedish iron miners exposed to low doses of radon daughters. *N. Engl. J. Med.*, 310, 1485-1494

Registrar General/Office of Population Censuses and Surveys. (1938, 1958, 1971, 1978) The Registrar General's Decennial Supplement for England and Wales. Occupational Mortality. 1931; 1951; 1961; 1970-72, London, Her Majesty's Stationery Office

Rinsky, R.A., Smith, A.B., Hornung, R., Filloon, T.G., Young, R.J., Okun, A.H. & Landrigan, P.J. (1987) Benzene and leukaemia. An epidemiologic risk assessment. *New Engl. J. Med.*, *316*, 1044-1050

Rinsky, R.A., Young, R.J. & Smith, A.B. (1981) Leukaemia in benzene workers. *Am. J. Ind. Med.*, *2*, 217-245

Roberts, R.S., Julian, J.A., Muir, D.C.F. & Shannon, H.S. (1984) Cancer mortality associated with the high-temperature oxidation of nickel subsulfide. In: Sunderman, F.W., Jr, ed., *Nickel in the Human Environment* (IARC Scientific Publications No. 53), Lyon, International Agency for Research on Cancer, pp. 23-35

Sandström, A.I.M., Wall, S.G.I. & Taube, A. (1989) Cancer incidence and mortality among Swedish smelter workers. *Br. J. Ind. Med.*, *46*, 82-89

Saracci, R. (1977) Asbestos and lung cancer: an analysis of the epidemiological evidence on the asbestos-smoking interaction. *Int. J. Cancer*, *20*, 323-331

Searle, C.E. (1976) Preface. In: Searle, C.E., ed., *Chemical Carcinogens*, (ACS Monograph No. 173), Washington, DC, American Chemical Society, pp. ix-xxvii

Selikoff, I.J., Hammond, E.C. & Seidman, H. (1980) Latency of asbestos disease among insulation workers in the United States and Canada. *Cancer*, *46*, 2736-2740

Seltser, R. & Sartwell, P.E. (1965) The influence of occupational exposure to radiation on the mortality of American radiologists and other medical specialists. *Am. J. Epidemiol.*, *81*, 2-22

Smith, P.G. & Doll, R. (1981) Mortality from cancer and all causes among British radiologists. *Br. J. Radiol.*, *54*, 187-194

Taylor, F.H. (1966) The relationship of mortality and duration of employment as reflected by a cohort of chromate workers. *Am. J. Pub. Health*, *56*, 218-229

Thomas, D.C. (1987) Pitfalls in the analysis of exposure-time-response relationships (with Appendix by Saracci, R. & Johnson, E.) *J. Chron. Dis.*, *40*, Suppl. 2, 71S-78S

Tokudome, S. & Kuratsune, M. (1976) A cohort study on mortality from cancer and other causes among workers at a metal refinery. *Int. J. Cancer*, *17*, 310-317

Tola, S., Kilpiö, J. & Virtamo, M. (1979) Urinary and plasma concentrations of nickel as indicators of exposure to nickel in an electroplating shop. *J. Occup. Med.*, *21*, 184-188

Torjussen, W. & Andersen, I. (1979) Nickel concentrations in nasal mucosa, plasma, and urine in active and retired nickel workers. *Ann. Clin. Lab. Sci.*, *9*, 289-298

Wada, S., Miyanishi, M., Nishimoto, Y., Kambe, S. & Miller, R.W. (1968) Mustard gas as a cause of respiratory neoplasia in man. *Lancet*, *i*, 1161-1163

Waldron, H.A. (1983) A brief history of scrotal cancer. *Br. J. Ind. Med.*, *40*, 390-401

Waldron, H.A., Waterhouse, J.A.H. & Tessema, N. (1984) Scrotal cancer in the West Midlands 1936-76. *Br. J. Ind. Med.*, *41*, 437-444

Walker, A.M. (1984) Declining relative risks for lung cancer after cessation of asbestos exposure. *J. Occup. Med.*, *26*, 422-426

Waterhouse, J.A.H. (1972) Lung cancer and gastro-intestinal cancer in mineral oil workers. *Ann. Occup. Hyg.*, *15*, 43-44

Weil, C.S., Smyth, H.F., Jr, & Nale, T.W. (1952) Quest for a suspected industrial carcinogen. *AMA Arch. Indust. Hyg. & Occup. Med.*, *5*, 535-547

CHANGES IN TOBACCO CONSUMPTION AND LUNG CANCER RISK: EVIDENCE FROM NATIONAL STATISTICS

A.D. Lopez

Global Epidemiological Surveillance and Health Situation Assessment, World Health Organization, Geneva, Switzerland

Introduction

Numerous epidemiological studies have been carried out to ascertain the role of tobacco smoking (particularly in the form of cigarettes) as a causal agent for lung cancer. It is now generally accepted that cigarette smoking is the major cause of lung cancer, accounting for at least 80% of all lung cancer deaths. This conclusion has emerged based on retrospective and prospective studies of specific population groups over the last 40 years or so (for an excellent review of the literature, see Doll and Peto (1981); US Public Health Service (1982)). If this association is indeed so strong, then it is logical to expect that the level and pattern of national lung cancer mortality rates would be broadly consistent with the observed patterns (at least in the past) of cigarette consumption. This paper examines to what extent this consistency is apparent from data available to the World Health Organization (WHO).

Before investigating the association between smoking and lung cancer, we shall consider the nature of the data available for such analysis in order to be quite clear about its limitations for this type of study.

Mortality data

Each year, about 70 countries report national data on causes of death to WHO. This information is provided either according to the 3- or 4-digit rubrics of the International Classification of Diseases (ICD) or according to the Short Lists developed in conjunction with each Revision of the ICD. For several countries, the data extend back to 1950, with 1988 being the latest year for which data are available. The statistics on causes of death are reported separately for males and females and by the conventional five-year age groups.

One of the major limitations of these data is the lack of information on associated factors as well as their heterogeneity. The statistics do not distinguish between smoking status of decedents, nor do they provide a breakdown of deaths by other variables, which may be useful for descriptive epidemiology, such as socioeconomic characteristics or residential history. Differences in diagnostic practices, coding procedures as well as medical education and use of diagnostic aids will also affect the comparability of causes-of-death statistics within and among countries. Successive revisions to the ICD have had minimal impact on the comparability of time series of lung cancer data and would appear to be of much lesser consequence than for some of the circulatory disorders.

Data on tobacco consumption

In order to evaluate the usefulness of information on tobacco consumption in relation to national trends in lung cancer, it is necessary to consider what data are available on 'smoking parameters' which influence exposure to tobacco carcinogens. These parameters include duration of smoking, inhalation patterns, age of initiation, as well as tobacco and tar content of cigarettes smoked. The parameters assume that all tobacco consumption has been in the form of manufactured cigarettes. If data on weight of tobacco (all forms) alone is available, then there is the additional problem of attributing the proportion of tobacco used for smoking cigarettes. In some countries where hand-rolled cigarettes are widely consumed (e.g., Australia and Norway), this form of tobacco consumption obviously would also affect exposure to risk of developing lung cancer.

Time series of data on tobacco consumption according to these parameters are, to the author's knowledge, unavailable. Moreover, tobacco consumption statistics are only rarely found for males and females separately although separate mortality time series for the sexes are available. Even where consumption data are provided by sex, the fact that females generally began to smoke in large numbers several decades after males implies that their exposure has been different from that of men and consequently so has their risk of developing lung cancer. In the United States, for example, it has been estimated that women's cumulative exposure to tobacco carcinogens has only been about 40% of that of men (Walker & Brin, 1988).

The existence of a latent period between exposure to cigarette smoking and dying from lung cancer implies that, in order to better assess current trends in lung cancer, specific information on smoking patterns by age in the past is needed, particularly for young adults. This information is not available for most countries although the smoking history of cohorts could be estimated from whatever time series data are available on smoking prevalence by age and sex.

Given these considerations, it is obviously extremely difficult (and may, in fact, be misleading in some circumstances) to compare national trends in overall cigarette consumption with the change in lung cancer mortality. It is even less appropriate to draw conclusions about the smoking–lung cancer relationship on the basis of cross-sectional data from the same (e.g. most recent) period. Nonetheless, surveys from several countries have pointed to a marked increase in cigarette smoking among women which is consistent with the rapid rise in female lung cancer mortality now observed in almost all developed countries. Hence, there is some justification for comparing national statistics on lung cancer and cigarette smoking although the results of such analyses must be interpreted with great caution in view of the confounding effects of several variables. Rather than examine changes in tobacco consumption (for which the data are less precise and comprehensive) in relation to trends in lung cancer (where the data are of better quality), it would seem more promising to review national trends in lung cancer and then examine to what extent these are consistent with available data on national tobacco consumption.

Overview of mortality trends from lung cancer[1]

Trends in age-standardized mortality rates

In view of the limited availability of data on tobacco consumption in developing countries, and the uncertain reliability and comparability of cause-of-death statistics in these countries, at least during the earlier part of the post-war period, the analysis will be restricted to those developed countries for which a sufficiently long time series of mortality data are available. These countries are listed in Tables 1A and 1B and are grouped according to the similarity of their lung cancer mortality patterns for males (Table 1A) and females (Table 1B). The classification is based on levels and trends of the age-standardized mortality rate, used here as a summary index of mortality over all ages.

For *males*, five groups of countries can be identified. The countries in Group 1 show a rapid rise in lung cancer mortality over the period 1950–84, with little evidence of a deceleration in the overall rate of increase. This group is in fact quite heterogeneous and includes countries of North America as well as countries of Europe. Countries in Groups 2 and 3 are also characterized by a steady rise in lung cancer death rates for males. These countries are distinguished from those in Group 1 by the *level* of mortality: thus, whereas the age-standardized mortality rates for males in Group 1 countries exceed 80 per 100 000 (and in some cases, e.g., Belgium, Netherlands, Czechoslovakia and Hungary exceed 100 per 100 000), the level for Group 2 nations (France, Ireland and Spain) is of the order of 65 to 80 deaths per 100 000 and for Group 3 (Japan, Norway and Portugal), between 35 and 45 per 100 000. What is common to all countries in these three groups, however, is that male mortality, as measured by the age-standardized death rate, is still increasing.

The trend characteristics for countries in Groups 4 (Federal Republic of Germany, Switzerland, Australia and New Zealand) and 5 (Austria, Finland, Sweden, and UK/England and Wales) are more distinctive. For the first of these groups, there is clear evidence of a *stagnation* of male lung cancer mortality around 70 to 72 deaths per 100 000 which has been apparent for the last decade or so. One may reasonably expect that death rates in these countries will soon begin to decline[2] at least in some age groups. This will be examined in greater

---

[1] Throughout this paper, the term 'lung cancer' actually refers to mortality from malignant neoplasm of the trachea, bronchus and lung (rubric 162 in ICD-8 and ICD-9 and rubrics 162, 163 in ICD-6 and ICD-7).

[2] This is already evident for New Zealand in the 1980s.

Table 1A. Patterns and trends of lung cancer mortality among males in developed countries, 1950-86

Country population	Age-standardized[a] mortality rate from lung cancer per 100 000					
	1950-54	1960-64	1970-74	1975-79	1980-84	1985-86[b]
Group 1. Mortality high and rising						
Belgium	43.5[c]	60.0	93.7	108.7	117.7	115.7
Canada	23.8	41.1	64.1	73.3	81.6	82.2
Czechoslovakia	55.0[c]	74.4	92.1	98.4	103.2	105.0
Denmark	26.0	46.8	66.1	72.8	80.9	81.3
Hungary	33.4[c]	43.7	63.5	76.3	91.5	104.9
Italy	16.0	34.0	57.7	68.7	80.1	86.0
Netherlands	37.8	67.6	99.8	111.9	116.9	117.4
USA	29.3	48.9	71.1	78.3	83.3	84.1
Group 2. Mortality intermediate and rising						
France	18.2	34.5	47.7	57.1	62.9	65.5
Ireland	19.8	38.5	59.4	68.0	71.8	77.4
Spain	13.0	25.5	35.5	45.6	52.5	58.9
Group 3. Mortality low but rising						
Japan	5.0	16.2	26.1	32.7	39.9	43.1
Norway	10.9	18.5	28.2	34.9	40.8	45.8
Portugal	11.2[c]	13.6	20.2	25.5	30.2	35.2
Group 4. Mortality now on plateau						
Australia	23.7	45.1	66.5	70.1	72.5	72.2
FR Germany	33.4	54.9	67.0	72.2	73.5	72.6
New Zealand	29.2	48.3	67.2	74.2	75.2	69.4
Switzerland	35.2	47.7	63.5	71.1	72.0	72.2
Group 5. Mortality declining						
Austria	65.9	71.9	75.6	75.5	73.6	69.7
Finland	62.4	85.7	97.1	97.3	92.3	80.2
UK/England & Wales	62.1	94.3	109.0	108.6	102.3	98.8
Sweden	13.4	22.4	32.8	38.0	36.5	35.0

[a] The 'European' population age structure was used as the standard
[b] Latest available year
[c] 1955-59

detail in a subsequent section. A *decline* in mortality is in fact already apparent for males in Group 5 countries. Sweden, for example, had a much lower level of mortality (13 per 100 000) from lung cancer in the early 1950s than the other three countries and the decline in mortality has occurred from a much lower peak

Table 1B. Patterns and trends of lung cancer mortality among females in developed countries, 1950-86

Country	Age-standardized[a] mortality rate from lung cancer per 100 000 population					
	1950-54	1960-64	1970-74	1975-79	1980-84	1985-86[b]
Group 1. Mortality high and rising						
Australia	4.5	5.6	9.7	12.2	15.1	16.6
Canada	4.8	6.0	11.0	15.6	21.5	26.1
Denmark	6.1	8.7	13.7	18.2	25.0	28.5
UK/England & Wales	8.8	12.5	19.2	22.8	26.3	28.1
Hungary	8.1[c]	10.1	12.0	13.1	15.7	18.0
Ireland	6.0	9.2	16.7	20.4	24.7	25.2
New Zealand	4.2	7.1	13.9	16.7	19.9	21.8
USA	5.7	7.4	15.1	20.4	26.9	30.6
Group 2. Mortality intermediate and rising						
Belgium	5.2[c]	6.1	7.4	8.5	9.4	11.1
FR Germany	5.6	7.0	7.1	7.9	9.2	10.6
Italy	3.9	5.8	7.1	8.2	9.1	10.1
Japan	1.9	5.8	8.0	9.5	11.2	11.9
Netherlands	4.7	5.1	5.9	7.3	10.1	12.6
Group 3. Mortality low but rising						
Austria	8.8[c]	8.4	9.5	10.2	11.8	11.9
Czechoslovakia	7.9[c]	8.1	8.1	9.1	10.2	11.0
Finland	6.1	5.4	6.1	7.5	8.6	9.2
Norway	3.7	3.9	5.6	6.9	9.5	12.6
Sweden	5.4	5.6	7.9	9.3	11.3	12.9
Switzerland	4.6	4.5	5.7	6.6	8.4	9.6
Group 4. Mortality now on plateau						
France	4.5	5.2	5.1	5.3	5.9	6.4
Portugal	3.1[c]	3.4	4.6	4.6	5.4	6.0
Spain	3.4	5.2	5.9	5.8	5.6	5.5

[a] The 'European' population age structure was used as the standard
[b] Latest available year
[c] 1955-59

level (i.e. at about 38 deaths per 100 000 which is only one-third to one-half of the peak observed for the remainder). In Austria, death rates increased only marginally between 1950–54 and 1970–74, whereas in England and Wales, the rise in mortality over the same period was of the order of 75%. The extent of the decline has been more marked in Finland, where male lung cancer mortality has fallen by roughly 17% or twice the decline observed for the other members of this group.

Among *females*, the predominant trend since 1950 has been a rapid rise in lung cancer mortality from a much lower base level than for males. Nonetheless, one may distinguish between four groups of countries largely on the basis of the observed level of mortality. Group 1 countries (primarily the Anglo-Saxon nations as well as Denmark and Hungary) all show a comparatively high mortality rate from lung cancer, ranging from 17–18 per 100 000 in Australia and Hungary to 28–30 per 100 000 in the USA, Denmark and England and Wales. A second group of countries (Group 2) with rising lung cancer mortality rates since 1950 can also be identified but which differ from those in Group 1 in that the current level of mortality (around 12 per 100 000) is considerably lower than for Group 1 countries. The countries of Group 3 also exhibit a level of mortality of around 10–12 per 100 000 but differ from Group 2 countries in that the rise in lung cancer mortality appears to have been delayed until the 1970s. For only three countries (France, Portugal and Spain (Group 4)) is the evidence of a rise in female lung cancer mortality less convincing in view of the comparatively low level of mortality (typically about 5 per 100 000) from the disease in these countries.

From this brief overview, it is clear that not only do the *levels* of mortality from lung cancer vary markedly between males and females, but also the characteristics of male and female patterns differ among countries, leading to the quite dissimilar categorization of countries in Tables 1A and 1B. These findings further reinforce the need to distinguish between the sexes when investigating the relationship between tobacco smoking and health consequences, such as lung cancer.

Age-specific trends

A summary index of mortality such as the age-standardized death rate is useful to indicate broad features of a country's mortality experience, particularly with regard to levels and trends. While this facilitates international comparisons and categorizations such as those in Table 1, summary indices of mortality conceal the often substantial variations in mortality patterns and trends experienced by specific age-groups. Moreover, in the situation of lung cancer for which a major generational effect is apparent, the age-standardized death rate combines the experience of several different generations of people, each of which has had very different exposure to the main causal agent – cigarette smoking – at different stages of life. In order to obtain a better appreciation of national lung cancer trends, it is necessary to look at the pattern by age and by cohort.

Trends in age-specific death rates from lung cancer for selected coutries are presented in Figure 1. While these countries are not necessarily representative of developments within each of the groups defined in Table 1, they are nonetheless fairly indicative of the various trends observed in developed countries over the last 30 years or so. To begin with, it is interesting to observe the trends for men in those countries (Finland, the United Kingdom) where overall death rates have started to decline. In these two countries, death rates at all ages 35-74[3] years have begun to fall, with the extent of the decline being greatest for the

---

[3]Only death rates up to age 75 have been considered in view of the probable diagnostic difficulties at older ages when multiple pathologies around the time of death are more frequent.

Figure 1. Age-specific trends in lung cancer mortality, by sex, selected countries, 1950-1985

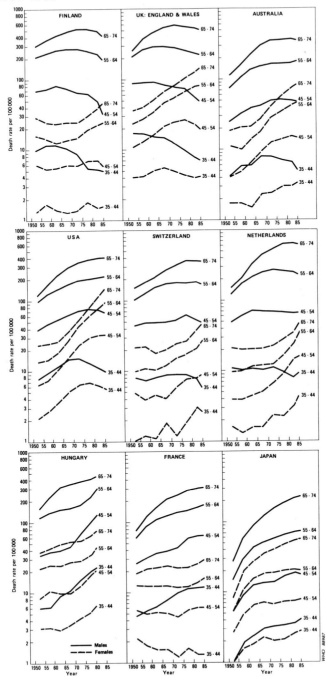

the younger age groups (below 55 years). In fact, the age pattern of declining death rates is remarkably uniform, with the decline commencing later and later with increasing age. This strongly suggests that the male cohorts with peak exposure to tobacco carcinogens have already disappeared in these countries and we are now witnessing declines in mortality due to a progressive lowering of risk. Consequently, the overall decline in male lung cancer mortality can be expected to continue, and even to accelerate as lower-risk younger adults move into the higher ages in these countries.

A somewhat similar pattern can be seen for Australia, the USA and, to a lesser extent, Switzerland, where death rates for younger males have begun to decline (particularly at ages 35–44). However, for these countries, there is still no evidence of a reduction in death rates above age 55, although there is clear indication that a *plateau* in rates has been reached and one may expect that death rates among older males will also begin to decline shortly. This accounts for the deceleration in the rate of increase of the age-standardized mortality rate and consequently a decline in the overall mortality from lung cancer among males in these countries can safely be predicted within the next few years.

The trend in age-specific death rates for men observed in the Netherlands falls somewhere between these two types. On the one hand, the decline in mortality among men aged 45–54 has been very slight with the level remaining more or less constant since the early 1960s. The same is true for men aged 35–44, although in this case there has been some improvement since the early 1970s. Given the relatively small numbers of deaths at these ages, even this recent development must be viewed with some reservation. At the higher ages, however, death rates have clearly peaked and have begun their descent. This will, no doubt, lead to a decline in the overall lung cancer mortality rate in the Netherlands which justifies its inclusion in the same group as the USA.

For the three remaining countries shown in Figure 1 (Hungary, France and Japan), death rates for males are still rising, most notably in Hungary. There is some evidence that the rate of increase in lung cancer mortality is slowing among older Japanese and French men, compared with rates in the 1950s, although a stagnation and eventual decline in mortality in these countries would appear to be improbable for several years yet.

Among females, the characteristics of national lung cancer trends are somewhat different. The only clear evidence of a decline in death rates is apparent for England and Wales at ages 35–54, where the fall in death rates first began among younger women, and for the USA at ages 35–44. There is also some indication that lung cancer death rates among women aged 45–54 in the USA have reached their peak but this is not so for older women among whom mortality in both countries is still rising rapidly. This trend is consistent with a marked lowering of risk among younger women in recent decades compared with older females whose exposure to risk (in part due to higher tar yield of cigarettes) has been considerably greater. Because of the overwhelming significance of death rates at the higher ages in the calculation of summary indices of mortality, there is, however, no immediate prospect of a decline in female lung cancer mortality in either the UK/England and Wales or the USA.

Interestingly, in Australia, Switzerland and the Netherlands, for which male lung cancer death rates at ages 35-54 have begun to decline, mortality among females at all ages, including younger women, is rising rapidly. One would, therefore, expect that the risk exposure of younger women in these countries has been quite different from that of counterparts in the USA and England. In Australia, for example, the prevalence of smoking among women at ages 16–44 years has either remained relatively stable during the period 1974–86, or, at the ages of peak prevalence (20–29 years), has actually risen slightly from 37.5% in 1974 to just over 40% in 1986 (Hill et al., 1988).

Female mortality from lung cancer in Japan is increasing in all age groups, although with the exception of the group 65–74 years where the rate of increase has slowed somewhat in recent years. Quite a different pattern is apparent for Finland and France (members of Groups 3 and 4 respectively) where lung cancer rates among younger women have been relatively constant but death rates of older females have been rising in recent years. In Hungary, on the other hand, lung cancer is rising sharply among women of all ages, paralleling the trend for men.

Cohort trends

While time trends in period-specific measures of mortality are useful for a more general assessment of changes in the health situation, causes of death such as lung cancer for which the exposure to a major causal agent differs for successive generations should equally be examined from a cohort point of view. Tables 2A and 2B present cohort death rates by age, for males and females respectively, for four selected countries where overall male mortality has either started to decline (Finland, UK/England and Wales) or will decline shortly (USA, the Netherlands).[4] In each case, one would expect that the death rates for recent generations, for whom exposure to tobacco carcinogens has been less than previous cohorts, should be somewhat lower than for cohorts born earlier in the century.

For both Finland and England and Wales, the trend towards lower mortality for recent cohorts of males is quite clear. Death rates among Finnish men aged 35–44 peaked for the cohort born around 1919–20, while for higher ages the maximum lung cancer mortality rate occurred among men born between about 1905 and 1915. Since then, death rates have progressively declined. Similarly, among British males aged 35–44, the peak in mortality from lung cancer occurred for the generation born around 1914–15, with death rates subsequently declining for each successive cohort. At higher ages, death rates reached their maximum for cohorts born 10 to 15 years earlier, reflecting their greater exposure to cigarette smoke.

In the United States of America, on the other hand, one needs to follow more recent birth cohorts in order to ascertain evidence of a decline in lung cancer

---

[4]For a more detailed compilation of cause-specific cohort mortality rates, see World Health Organization (1987)

Table 2A. Age-specific cohort mortality trends from lung cancer, males, selected countries

Central years of birth cohort	Age-specific death rates per 100 000 population								
	30-34	35-39	40-44	45-49	50-54	55-59	60-64	65-69	70-74
Finland									
1899-1900					96.9	192.5	330.4	468.0	582.8
1904-05				42.9	103.7	210.1	352.6	494.0	621.5
1909-10			14.6	45.5	111.9	219.7	345.9	470.9	578.4
1914-15		4.8	17.6	51.5	102.8	209.6	324.7	423.5	
1919-20	0.8	5.6	18.8	49.3	97.1	186.0	307.9		
1924-25	1.4	5.5	17.1	40.8	97.8	168.5			
1929-30	1.3	4.7	13.8	32.7	72.6				
UK/England & Wales									
1899-1900					119.8	229.6	364.5	527.2	681.3
1904-05				58.2	125.1	231.5	373.8	522.0	668.8
1909-10			24.8	59.3	123.4	221.9	359.0	499.8	638.5
1914-15		9.8	25.5	56.6	116.9	212.7	335.3	459.6	
1919-20	3.7	9.3	22.5	53.1	107.1	189.5	300.2		
1924-25	3.4	8.9	21.5	50.5	102.7	172.1			
1929-30	3.5	7.8	18.7	41.5	79.3				
1934-35	2.5	6.0	14.0	32.6					
1939-40	2.4	5.6	11.9						
1944-45	1.7	3.9							
1949-50	1.2								
Netherlands									
1899-1900					66.7	144.3	267.6	430.0	586.2
1904-05				33.2	82.9	174.1	318.1	505.8	725.9
1909-10			15.6	39.4	98.7	194.8	348.3	554.4	771.6
1914-15		6.3	15.2	44.1	98.7	204.1	350.5	539.2	
1919-20	1.9	5.7	16.8	42.8	99.0	190.3	333.9		
1924-25	2.4	5.1	15.6	40.4	90.3	179.1			
1929-30	1.8	5.4	17.1	43.5	92.3				
USA									
1899-1900					51.7	104.6	180.0	280.1	377.3
1904-05				26.6	63.0	121.6	211.9	317.8	434.5
1909-10			11.7	31.6	72.6	140.4	238.6	342.0	461.4
1914-15		4.3	13.6	36.2	83.1	158.3	257.3	368.6	
1919-20	1.6	5.4	16.8	42.9	91.0	162.7	262.8		
1924-25	2.0	6.6	20.6	50.2	100.2	172.9			
1929-30	2.4	7.9	22.2	51.0	99.1				
1934-35	2.3	8.1	20.4	47.1					
1939-40	1.9	6.4	17.7						
1944-45	1.6	5.4							
1949-50	1.2								

Table 2B. Age-specific cohort mortality trends from lung cancer, females, selected countries

Central years of birth cohort	Age-specific death rates per 100 000 population								
	30-34	35-39	40-44	45-49	50-54	55-59	60-64	65-69	70-74
Finland									
1899-1900					7.4	11.5	14.1	22.8	30.3
1904-05				5.0	6.0	10.9	15.7	20.7	34.9
1909-10			2.6	4.0	6.2	12.1	17.9	26.5	43.1
1914-15		0.7	2.1	4.9	6.5	12.0	25.0	35.7	
1919-20	0.7	1.4	1.4	5.7	7.3	15.0	29.5		
1924-25	-	1.3	2.0	4.7	9.6	16.2			
1929-30	-	0.7	2.1	4.9	9.9				
UK/England & Wales									
1899-1900					13.5	23.9	40.7	65.1	85.4
1904-05				8.4	16.4	29.0	49.4	77.0	105.9
1909-10			5.2	10.0	20.1	38.1	63.7	96.5	133.7
1914-15		2.7	5.6	12.9	27.1	47.1	81.2	121.9	
1919-20	1.5	3.0	6.8	15.4	31.5	54.7	93.3		
1924-25	1.5	3.2	7.7	17.7	36.1	63.0			
1929-30	1.2	3.2	6.9	16.5	31.6				
1934-35	1.1	2.8	6.3	13.9					
1939-40	0.8	2.3	5.9						
1944-45	0.8	2.4							
1949-50	0.7								
Netherlands									
1899-1900					4.6	8.1	13.8	18.0	26.7
1904-05				3.2	4.9	9.6	14.5	19.5	32.6
1909-10			2.1	3.0	5.2	9.7	14.9	24.1	38.1
1914-15		1.2	1.8	3.8	6.3	10.4	17.1	33.2	
1919-20	0.5	0.8	1.9	4.2	8.5	16.4	28.3		
1924-25	0.5	1.0	1.8	5.0	11.2	21.2			
1929-30	0.8	1.3	2.6	6.9	18.7				
USA									
1899-1900					7.7	12.3	20.5	35.2	54.5
1904-05				5.0	9.2	16.4	29.4	50.9	79.8
1909-10			2.9	6.1	12.7	25.8	47.3	76.3	120.2
1914-15		1.4	3.6	8.7	19.4	40.8	70.5	114.7	
1919-20	0.7	1.8	5.2	13.0	28.9	53.2	92.7		
1924-25	0.8	2.3	7.1	18.2	37.1	68.4			
1929-30	0.8	3.0	9.1	22.4	43.9				
1934-35	0.9	3.6	10.6	24.0					
1939-40	1.0	3.6	9.9						
1944-45	0.9	3.4							
1949-50	0.8								

mortality. Thus, death rates at ages 35–44 have apparently begun to decline with the cohort of males born around 1939–40, which is consistent with the much more recent and modest decline in period-specific death rates in the USA compared with Finland and the UK/England and Wales (see Figure 1). Furthermore, contrary to the experience of these two countries, male lung cancer death rates among older adults have continued to rise for successive cohorts, although the rate of increase at most ages has fallen considerably for more recent cohorts. For example, at ages 45–54, the ratio of mortality for successive generations of men is as follows:

Age group	Ratios of death rates for cohorts born around				
	1914–15 / 1909–10	1919–20 / 1914–15	1924–25 / 1919–20	1929–30 / 1924–25	1934–35 / 1929–30
45–49	1.43	1.49	1.40	1.23	1.07
50–54	1.53	1.49	1.28	1.18	–

In other words, between the cohorts born around 1929–30 and 1924–25 there was a 23% increase in mortality at ages 45–49, whereas for those born in 1934–35, lung cancer at these ages was only 7% higher than for the 1929–30 birth cohort.

The pattern for the Netherlands is characterized by a general stability of age-specific mortality rates from one generation to the other, at least at the younger ages, although there is some evidence (particularly among men in their 50s) that death rates for more recent birth cohorts are declining. Once again, this trend is entirely consistent with what emerged from the period-specific analysis for men in the Netherlands.

Similarly, the cohort trends for females are very much what one would expect from a glance at the period-specific patterns. Thus, in the USA and England and Wales, where lung cancer death rates have now begun to fall among younger women, one can observe the peak in cohort mortality for British women born around 1924–25, while for their American counterparts, the most affected cohort were those women born five years later. Among older women (i.e., those aged 55 years or more), on the other hand, cohort death rates are still rising.

As one might expect from the period analysis, lung cancer among Finnish women born in recent years has remained relatively stable at younger ages (below about 49 years) but has steadily increased with successive cohorts at older ages. Also, among Dutch women, cohort death rates are rising at all ages, although the relatively small numbers of deaths at younger adult ages, as for some other countries, makes the interpretation of trends rather more difficult.

Relationship with national patterns of tobacco consumption

Before examining the interrelationships between these trends in lung cancer mortality and tobacco consumption, it is perhaps appropriate to recall the limitations of the two data sets available for this purpose. Both are aggregated statistics reflecting the experience of rather different (with regard to health behaviours) population subgroups whose risk exposure has varied enormously, not only within a country at the same point in time (e.g., between males and females) but also within a given population subgroup over time (e.g., the markedly lower tar yields compared with previous years).[5] The limited availability of comparable (in terms of coverage and the methodology used to obtain the estimates) data on tobacco consumption by age and sex, as well as the smoking characteristics of smokers, particularly for earlier years, is also a major impediment for this type of analysis.

Figure 2. International correlation between overall consumption of manufactured cigarettes per adult[a] in 1960 and age-standardized death rate (both sexes combined from lung cancer around 1985

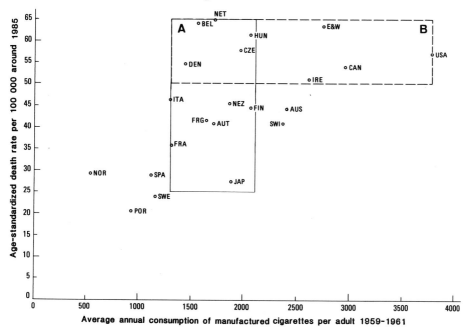

[a] Persons aged 15 years and over

As a first step, having excluded *a priori* that a relationship should exist between *current* levels of tobacco consumption and *current* lung cancer mortality rates in view of the latent period expected between dose (tobacco consumption) and response (death from lung cancer), nonetheless, some correlation between current mortality levels and levels of exposure some 20 to 30 years ago can be expected. As Figure 2 illustrates, this association is only broadly apparent. In

---

[5]For example, compared with 1948, the average tar yield per cigarette in Britain had fallen 9% by the mid-1950s, declining by a further 14% by the mid-1960s and then again by 27% by the mid-1970s (Doll & Peto, 1981, p. 1299)

particular, there is a group of countries (Box A) for which the level of mortality from lung cancer varies considerably despite the fact that their per capita adult consumption of manufactured cigarettes around 1960 was reasonably similar. Conversely, countries in Box B have rather similar levels of lung cancer although the per capita consumption varies from less than 1500 cigarettes per adult per year in Denmark to over 3500 cigarettes in the USA.

On reflection, however, this is perhaps to be expected in view of the problems with the data. Manufactured cigarettes are only one (albeit major) means of consuming tobacco, and in some countries, such as Norway, hand-rolled cigarettes account for a very high proportion of all cigarettes smoked (Rothwell et al., 1987). Moreover, per capita consumption of manufactured cigarettes gives no indication of what proportion of the population smoke, how much is smoked and in what fashion. Similarly, the age-standardized death rate merely indicates the *overall* level of mortality from lung cancer and gives no indication of the age-specific level. Doll and Peto (1981) have proposed a further refinement of this risk–outcome relationship by focusing on cigarette consumption in early adult life, a major determinant of lung cancer risks at higher ages based on data for US veterans reported by Kahn (1966). The authors found a 'reasonable correlation' between manufactured cigarette consumption in 1950 when a given cohort was entering adult life and the subsequent lung cancer mortality of that group upon entering middle-age (Figure 3). Nonetheless, even this correlation, as the authors freely admit, is to some extent fortuitous in view of the 'international differences in cigarette consumption, puff frequency, style of inhalation, butt length, additional use of non-manufactured cigarettes (and other forms of tobacco), and national consumption of cigarettes in the intervening years between 1950 and 1975' (Doll & Peto, 1981, p. 1301).

Additionally, the use of total adult consumption as a proxy for the smoking patterns of young adults further confounds the interpretation of this relationship, although this to some extent may have been offset by smoking behaviour before the mid-1970s.

Some further, although rather imprecise, evidence of an association between national tobacco consumption levels and lung cancer mortality rates can be gleaned from an examination of sex-specific data. It is well known that women, for a variety of reasons including the social acceptability of female smoking, began to smoke much later than males and have generally incurred a much lower exposure to risk due to differences in consumption patterns compared with males. This *a priori* should imply a lower level of lung cancer mortality which is precisely what has been observed. Table 3 summarizes some of the relevant data for selected developed countries.

Male mortality from lung cancer is between 3 and 10 times the level for females in these countries. This is at least consistent with the observation that, for every country shown in the table, not only was smoking more common among males, but also among smokers, males consumed more than females. These data must, of course, be interpreted with considerable caution given the multiplicity of cohort experience summarized by the age-standardized death rate as well as the obvious differences in exposure indicators between countries. Even so, the relative magnitude of the sex-mortality ratio is broadly consistent with the relative male/female risk suggested by the prevalence and consumption data.

Figure 3. International correlation between manufactured cigarette consumption per adult in 1950 while one particular generation was entering adult life (in 1950) and lung cancer rates in that generation as it enters middle age (in the mid-1970s).

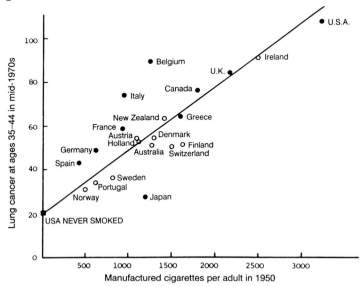

● Rates based on >100 deaths; ○ rates based on 25–100 deaths.
Reproduced from Doll & Peto (1981), by permission of Oxford University Press and the authors

For some countries, disaggregated statistics on smoking patterns are available for specific age groups or cohorts which can be used to provide a much more precise view of the relationship between tobacco consumption and mortality. Doll (1982), for example, has compared the trends since 1950 in lung cancer mortality with that of cigarettes smoked for men and women aged 40–49 years and 60–69 years in England and Wales. For younger males, death rates began to fall around 1960, which is consistent with the drop in tar which commenced immediately after the second world war. Among young women, however, for whom the death rate was lower at each level of consumption compared with males, the decline in tar content per cigarette was to a large extent countered by the increase in the number of cigarettes smoked. As a result, the net amount of tar consumed only began to fall around 1965 and, hence, the decline in female lung cancer death rates occurred much later (in the late 1970s) than for males.

The contrast between male and female patterns is even more obvious at higher ages (60–69 years). For men in this age group, the decline in mortality began some 10 to 15 years later than for those aged 40–49 years, corresponding to a ten-year delay (compared with younger males) in the onset of the decline in tar delivery. On the contrary, there has been no decline in the net tar consumption of older women among whom the rise in mortality from lung cancer shows no signs of abating.

Table 3. Lung cancer mortality in the mid-1980s and consumption indicators from earliest available data, by sex, in selected countries

Country	Age-standardized death rate from lung cancer			Prevalence of smokers		Indicator of consumption for males and females
	M	F	M/F Date	M	F	
Australia	72.2	16.6	4.3 1974	41	29	% smokers consuming > 25 cigs/day 1976: M = 25%, F = 15%
Canada	82.2	26.1	3.1 1966	54	32	Av. no. cigarettes/day/smoker: 1970: M = 12, F = 6
UK/England and Wales	98.8	28.1	3.5 1960	74	42	% smokers consuming > 20cigs/day: 1975: M = 41%, F = 26%
Finland	80.2	9.2	8.7 1967	50	15	Av. no. cigarettes/day/smoker: 1978: M = 18, F = 12
France	65.5	6.4	10.2 1953	77	35	% smokers consuming > 10cigs/day: 1976: M = 54%, F = 20%
Italy	83.5	9.6	8.7 1949	71	10	Av. no. cigarettes/day/smoker: 1949: M = 14, F = 8
Japan	43.1	11.9	3.6 1967	82	18	Av. no. cigarettes/day/smoker: 1964: M = 19, F = 12
Netherlands	117.4	12.6	9.3 1958:	90	29	Av. no. cigarettes/day/smoker: 1958: M = 18, F = 8
Norway	45.8	12.6	3.6 1956	65	23	Av. no. cigarettes/day/smoker: 1956: M = 13, F = 10
USA	84.1	30.6	2.7 1965	52	34	% smokers consuming > 25cigs/day: 1965: M = 24%, F = 13%

Source: Age-standardized death rates calculated from the WHO mortality data-bank; prevalence and consumption figures taken from Rothwell et al. (1987); Rothwell & Masironi (1986)

A very similar pattern can be observed for the United States. Among males, the maximum exposure to tobacco carcinogens occurred for the generations born between 1911 and 1920 (Devesa et al., 1987) whereas for women, the peak in smoking prevalence occurred for generations born between 1931 to 1940. This sex differential in exposure patterns is reflected in the differential mortality trends of men and women reported in the previous section, with male mortality (at least for the younger ages for which data are available) beginning to decline with the generations born since about 1935. On the contrary, female mortality is falling only in the very youngest age groups, corresponding to birth cohorts since 1940. Overall, lung cancer incidence rates for men peaked in the United States in 1982 (Horm & Kessler, 1986). Smoking among males, on the other hand, began to decline around the mid 1960s, with prevalence falling steadily from just over 50% in 1965 to 35% in 1983. These trends are consistent with a 20-year lag period between maximum exposure and maximum effect (lung cancer incidence). Smoking prevalence among women over the same period has hardly altered and both incidence and mortality are rising at an annual rate of about 6% (Horm & Kessler, 1986).

Certainly, the trends illustrated in Table 1 and Figure 1 are broadly consistent with what is known about changes in risk. Cigarette consumption among males has undoubtedly fallen in the USA, Canada, Finland and England and Wales which, coupled with the trend towards lower tar intake per cigarette, is consistent with the observed decrease in male death rates. The marked declines in male lung cancer mortality in Finland, for example, are a direct consequence of the considerable decrease in tobacco consumption in Finland since 1970, accompanied by dramatically lower tar yields (IARC, 1986). In contrast, consumption among males has risen in Japan (Rothwell & Masironi, 1986) and we are now witnessing a rapid rise in male lung cancer in this country.

A very impressive graph illustrating the similarity of the trends in lung cancer mortality (lagged by 20 years) with the trend in tobacco consumption from smoking has been prepared for Canada (Figure 4) and undoubtedly similar trends for other countries can be found. Interestingly, the graph suggests that overall lung cancer rates in Canada are about to decline (if the relationship between the two curves is to be maintained). This is supported by the age-sex-specific trend data which indicate that death rates have peaked for younger males and that the rate of increase among older men is rapidly declining (Table 4). Death rates for females are still rising at all ages but given the predominance of male deaths in overall mortality statistics for lung cancer, the national death rate appears set to decline.

Despite this close correlation between the two trends, one must not lose sight of the fact that aggregate statistics such as those plotted in the graph (Figure 4) confound the dose-response experience of several generations. Nonetheless, here is a good example of the use of national statistics to demonstrate broadly the relationship between smoking and lung cancer, a relationship which will be all the more apparent once lung cancer death rates in Canada can be shown to have begun to decline.

Concluding remarks

Perhaps the most interesting of the age-specific trends to follow in the future are the changes which have been observed in several countries among younger adults aged 35–44 years and even those aged 45–54. It is at these ages that the first signs of a decline in mortality from lung cancer have appeared. Assuming a 20-year latent period between massive exposure to cigarette smoking and the

Figure 4. Tobacco consumption from smoking[1], 1920–1985, and lung cancer mortality rates[2], Canada, 1940–1985, for both sexes combined

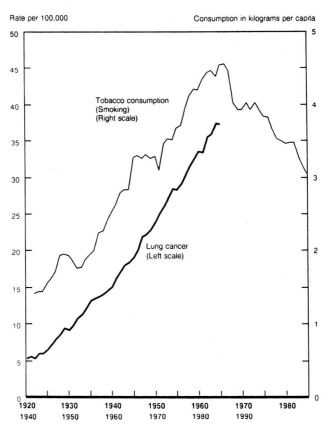

[1]Tobacco consumption from smoking was calculated in kilograms per capita (based on population 15+) and excludes snuff and chewing tobacco

[2]Mortality rates are age-standardized to the 1971 Canadian population, and include deaths for all age groups

Source: Health Division, Statistics Canada, and Surveillance and Risk Assessment Division, Health and Welfare Canada.

Reproduced with permission from: Canadian Cancer Society (1988) *Canadian Cancer Statistics*, Toronto

appearance of lung cancer, the fact that the decline in mortality commenced as early as 1950–54 for British males means that the changes in consumption patterns among adolescents as far back as 1930 or so need to be examined. To the author's knowledge, these data are not available and, hence, it is rather difficult to evaluate the age-specific trends depicted in Figure 1 in relation to changes in consumption patterns. Certainly, a decline in smoking prevalence among young men has been observed in several countries (e.g. in Canada, prevalence among males aged 15–19 fell progressively from 35% in 1966 to 20% in 1983, and from 60% to 37% among men aged 20–24 in both Canada and the USA (Rothwell & Masironi, 1986)), and these trends may be indicative of a longer-term decline commencing several years earlier. This is, at least, what could be expected from the decline in lung cancer mortality for young adult males in these countries which began in the early 1970s. At any rate, the observed decrease in prevalence since the mid-1960s, coupled with decreased tar intake, will ensure a continued decrease in lung cancer mortality for young men in these and other countries with a similar age-specific trend pattern.

Table 4. Ratios of age-specific mortality rates from lung cancer for successive quinquennia, Canadian males aged 35–74, 1950–85

Periods	Ratios of death rates at ages			
	35–44	45–54	55–64	65–74
1955-59/1950-54	1.00	1.14	1.39	1.43
1960-64/1955-59	1.23	1.17	1.19	1.35
1965-69/1960-64	1.33	1.18	1.16	1.33
1970-74/1965-69	1.20	1.19	1.18	1.21
1975-79/1970-74	0.87	1.08	1.10	1.12
1980-84/1975-79	0.95	1.02	1.10	1.11
1985/1980-1984	0.89	0.91	1.01	1.03

It is clearly indispensable for countries to monitor closely changes in risk exposure to tobacco carcinogens if they are to respond adequately to observed changes (both positive and negative) in lung cancer (and other effects causally associated with smoking). As a minimum requirement, comparable age-sex-specific data on tobacco consumption and exposure to risk should be systematically collected and used as a basic information support for health planning in countries where lung cancer is perceived as a major health problem. Wider dissemination of these data to the international scientific community would facilitate their use for comparative international analyses and would assist countries in evaluting progress towards their own health goals.

References

Canadian Cancer Society (1988) *Canadian Cancer Statistics 1988*, Toronto

Devesa, S.S., Silverman, D.T., Young, J.L., Pollack, E.S., Brown, C.C., Horm, J.W., Percy, C.L., Myers, M.H., McKay, F.W. & Fraumeni, J.F., Jr (1987) Cancer incidence and mortality trends among whites in the United States, 1947-84. *J. Natl Cancer Inst.*, 79, 704-770

Doll, R. (1982) Trends in mortality from lung cancer in women. In: Magnus, K., ed., *Trends in Cancer Incidence: Causes and Practical Implications*, New York, Hemisphere, pp. 223-230

Doll, R. & Peto, R. (1981) *The Causes of Cancer. Quantitative Estimates of Avoidable Risks of Cancer in the United States Today*, Oxford, Oxford University Press

Hill, D.J., White, V.M. & Gray, N.J. (1988) Measures of tobacco smoking in Australia 1974-1986 by means of a standard method. *Med. J. Australia*, *149*, 10-12

Horm, J.W. & Kessler, L.G. (1986) Falling rates of lung cancer in men in the United States. *Lancet*, *i*, 425-426

IARC (1986) *IARC Monographs on the Evaluation of the Carcinogenic Risk of Chemicals to Humans.* Vol. 38, *Tobacco Smoking,* Lyon, International Agency for Research on Cancer

Kahn, H.A. (1966) The Dorn study of smoking and mortality among US Veterans: Report on eight and one-half years of observation. In: Haenszel, W., ed., *Epidemiological Approaches to the Study of Cancer and Other Diseases* (National Cancer Institute Monographs No. 19), Bethesda, MD, US Public Health Service, pp. 1-125

Lee, P.N., ed. (1975) *Tobacco Consumption in Various Countries.* Research Paper No. 6, 4th ed., London, Tobacco Research Council

Rothwell, K. & Masironi, R. (1986) *Cigarette smoking in developed countries outside Europe.* WHO Document WHO/SMO/86.2

Rothwell, K., Masironi, R. & O'Byrne, D. (1987) *Smoking in Europe.* WHO Document WHO/SMO/EURO/87.1

US Public Health Service (1982) The health consequences of smoking: cancer. A report of the Surgeon General of the Public Health Service. US Department of Health and Human Services Office on Smoking and Health, Washington, DC, US Government Printing Office

Walker, W.J. & Brin, B.N. (1988) US lung cancer mortality and declining cigarette tobacco consumption. *J. Clin. Epidemiol.*, *41*, 179-185

WHO (1987) *World Health Statistics Annual*, Geneva, World Health Organization, pp. 350-411

CHANGES IN TOBACCO CONSUMPTION AND LUNG CANCER RISK: EVIDENCE FROM STUDIES OF INDIVIDUALS

D.R. Shopland

Division of Cancer Prevention and Control,

National Cancer Institute,

Bethesda, MD, USA

Smoking, especially cigarette smoking, is the single largest cause of cancer mortality in the United States and most of the developed world (US Department of Health, Education and Welfare, 1979). Smoking is an established cause of cancer at several sites, including the lung, larynx, oral cavity, pharynx and oesophagus, and is a contributing factor in the development of cancers of the bladder, kidney and pancreas (IARC, 1986; US Department of Health and Human Services, 1982). More recent data indicate that smoking is a probable cause of cancer of the cervix in women and may contribute to and result in death from stomach cancer in men and women (IARC, 1986). Overall, smoking is responsible for almost one third of all cancer deaths in the United States, most due to cancers of the respiratory system (Doll & Peto, 1981) (in the United States, lung cancer alone now accounts for more than 28% of all cancer deaths annually). This represents more than 150 000 of the approximately 500 000 cancer deaths estimated to have occurred in 1989 by the American Cancer Society (ACS) (American Cancer Society, 1989).

Every medical and public health institution that has conducted an objective review of the scientific evidence has reached the conclusion that cigarette smoking is the single largest cause of excess cancer morbidity and mortality for both men and women. Only the tobacco industry and those affiliated with growing or manufacturing tobacco products seriously question the scientific basis linking smoking and tobacco use to early mortality from cancer and other chronic diseases. Although the evidence linking smoking and cancer is well known and accepted among both the public and scientists, surprisingly little is known about the health benefits from quitting smoking in order to reduce these risks.

Because cigarette smoking contributes such a large burden to overall cancer deaths and rates, it is important from a public health perspective to examine the benefits of giving up smoking. If stopping smoking can reduce an individual's risk of developing the disease, this knowledge could have enormous implications for cancer prevention and control if large numbers of current smokers could be persuaded to quit. Of particular importance is how much time is required for a current smoker's risk to decline following cessation and whether this risk will approach or equal that of a non-smoker.

The purpose of this paper is to examine what is known about reducing the cancer risks associated with smoking, including a brief examination of data on the possible benefits of smoking low-tar and low-nicotine cigarettes. Sources of information are limited to major prospective and retrospective studies that contain sufficient numbers of ex-smokers to permit conclusions to be drawn. Most of these studies were initiated in the 1950s and early 1960s and included cohorts of individuals born before the turn of the century. During this period, males changed from other forms of tobacco use such as smokeless tobacco, pipes and cigars to cigarettes (Shopland, 1986), and many cohorts of women did not smoke in large numbers or took up smoking in their late twenties or early thirties (Harris, 1983; US Department of Health and Human Services, 1980). Such cohorts would be expected to have lower total lifetime smoking exposures compared with cohorts (particularly women) born after 1900, when cigarette smoking became more socially acceptable (US Department of Health and Human Services, 1980). An exception to this body of evidence is the new ACS Fifty State Study initiated in 1982. This study contains information on 1.2 million men and women who were followed for four years. Results for women are available in more detail than for men; however, data for both men and women will be discussed.

Although this paper will concentrate on the effects of stopping smoking on lung cancer risk, limited data on other cancer sites are included from the new ACS study. Where mortality information is lacking from the most recent follow-up period from these studies, earlier data will be cited.

The prospective cohort mortality studies

In the 1982 Surgeon General's report (US Department of Health and Human Services, 1982), eight major prospective studies were cited that examined the relationship between cigarette smoking and cancer of various sites. At that time, these studies cumulatively represent nearly 20 million person–years of observation and more than 300 000 deaths. At that time also, the ACS Fifty State Study was only in its initial stage of enrolling cases and was thus not available for review. In this paper, data related to the benefit of quitting smoking will be limited to four of the original eight prospective studies, the large case–control study of Lubin *et al.* (1984b), and preliminary data from the new ACS Fifty State Study (Garfinkel & Stellman, 1988). These studies were selected primarily because they contain large numbers of deaths; four of these studies also included women. The reader should refer to the individual references cited for a more complete description of each of the studies.

British Physicians' Study

More than 34 000 men and 6200 women physicians responded to a questionnaire distributed by the British Medical Association in 1951. With few exceptions, all physicians who replied in 1951 were followed to their deaths or for a minimum of 20 years (for males) and 22 years (for females). Data on changes in tobacco-use behaviour were collected at various intervals through 1973. More than 11 000 deaths from all causes occurred in this population (Doll & Peto, 1976; Doll et al., 1980).

American Cancer Society Twenty-five State Study

During 1959 to 1960, the ACS enrolled slightly more than one million men and women in their Cancer Prevention Study I (ACS Twenty-five State Study). This cohort was followed for a total of 12 years; nearly 93% of all survivors were successfully traced. Although not a representative sample of the US population, all segments of the population were included except groups that were difficult to trace. More than 150 000 deaths were recorded, including more than 2500 lung cancer deaths (Hammond, 1966; Hammond et al., 1977).

US Veterans' Study

This study followed the mortality experience of nearly 300 000 veterans who held Government life insurance policies in the 1950s. Almost all policy-holders were white males. More than 107 000 deaths occurred in this population during the first 16 years of follow-up. Twenty-six-year follow-up information has been collected, but not reported, on this cohort. Data in this paper will be limited to information published for the first 16 years. More than 3000 lung cancer deaths were recorded among this cohort (Kahn, 1966; Rogot & Murray, 1980).

Japanese study of 29 health districts

In 1965 265 000 Japanese men and women were enrolled in a prospective study in 29 health districts, representing between 90 and 99% of the total population 40 years and older. Follow-up information reported from this study has been sporadic. Data for women have been limited primarily to the first five to eight years of follow-up. Data for men have been reported through 13 years. More than 40 000 total deaths have occurred among the cohort, including more than 10 000 cancer deaths (Hirayama, 1974, 1977).

American Cancer Society Fifty State Study

In September 1982, the ACS initiated Cancer Prevention Study II in which 1.2 million men and women in all 50 states were enrolled in a prospective design. Initial results of this study are just now emerging. Currently, there is more complete information on women than men. By July 1988, death certificates had not been received for about 9% of male and 13% of female deaths. A total of 1006 lung cancer deaths among the cohort of 619 225 women were recorded (Garfinkel & Stellman, 1988).

Differences in lung cancer risk and smoking behaviour between men and women

Epidemiological research over 30 years has conclusively demonstrated that cigarette smoking is associated with a significantly elevated risk of developing and dying from lung cancer. Two major reviews examining this relationship have been published: *The Health Consequences of Smoking: Cancer. A Report of the Surgeon General* (US Department of Health and Human Services, 1982) and the 1986 *IARC Monograph on Evaluation of the Carcinogenic Risk of Chemicals to Humans: Tobacco Smoking* (IARC, 1986).

For men, the average lung cancer risk was about 10 to 12 times greater if they smoke than for men who have never smoked. Among women cigarette smokers, the lung cancer risk was also elevated, but the ratio between smokers and non-smokers was not as high as that for males. The differences between male and female smokers and non-smokers primarily reflected the observed historical differences in total lifetime smoking experiences between the sexes. Men began cigarette smoking earlier than women, they smoked more cigarettes per day, inhaled more deeply, had smoked for longer periods of time, and consumed cigarettes with a tar content higher than those smoked by women.

In the United States, more than half of all men were classified as regular smokers according to studies and surveys conducted over the previous 60 years. In fact, a majority of males were probably regular cigarette users by the end of the first world war or soon after (US Department of Health and Human Services, 1980). In comparison, only 18% of adult women were smokers in 1935 (Fortune Magazine, 1935). Even as late as 1955, twice as many men as women smoked (Haenszel et al., 1956); in the ten-year period between 1955 and 1965, male smoking remained stable at more than 50%. Female smoking rates during this period increased to around 34% – the highest smoking prevalence rate ever reported for adult females in the USA (US Department of Health and Human Services, 1980). Today, women smoke fewer cigarettes per day than men, and more men than women are heavy smokers (National Center for Health Statistics, 1986). When smoking behaviour is examined by individual birth cohorts, further differences between men and women are apparent. Some male cohorts have cigarette-use rates exceeding 70%, whereas no cohort of women is observed to have smoking rates higher than approximately 50% (Harris, 1983; Shopland, 1987). Furthermore, most women currently at risk for developing lung cancer did not begin smoking until well into their twenties or early thirties, whereas the vast majority of men began regular smoking during adolescence (McGinnis et al., 1987). Males also have a history of substantial use of pipes and cigars whereas few females ever smoked these products (US Department of Health and Human Services, 1980).

Cigarette dose and lung cancer risk

The risk of developing and dying from lung cancer is strongly dependent on the total lifetime dose of cigarette smoke. In the prospective studies, lung cancer mortality has been examined by number of cigarettes smoked per day, length of time one has smoked, age of initiation, inhalation characteristics, and tar and nicotine content of cigarettes used. Table 1 provides data for lung cancer risk overall by number of cigarettes smoked daily for males and females. Generally, the higher the daily consumption of cigarettes, the greater the risk. Lung cancer mortality ratios among heavy smoking males are 20 or more times higher than in non-smoking males. For females, the risks also increase with increasing numbers of cigarettes smoked; but even for women who smoke heavily (i.e., two or more packets daily), the risk for lung cancer is lower than that among men who smoke similar numbers of cigarettes.

Table 1. Lung cancer mortality ratios for men and women, by current number of cigarettes smoked per day – prospective studies

Population	Men		Women	
	Cigarettes smoked per day	Mortality ratios	Cigarettes smoked per day	Mortality ratios
ACS 25 State study	Non-smoker	1.00	Non-smoker	1.00
	All smokers	8.53	All smokers	3.58
	1–9	4.62	1–9	1.30
	10–19	8.62	10–19	2.40
	20–39	14.69	20–39	4.90
	40+	18.71	40+	7.50
British Doctors' study	Non-smoker	1.00	Non-smoker	1.00
	All smokers	14.00	All smokers	5.00
	1–14	7.80	1–14	1.28
	15–24	12.70	15–24	6.41
	25+	25.10	25+	29.71
US Veterans' study	Non-smoker	1.00		
	All smokers	11.28		
	1–9	3.89		
	10–20	9.63		
	21–39	16.70		
	> 40	23.70		
Japanese study	Non-smoker	1.00	Non-smoker	1.00
	All smokers	3.76	All smokers	2.03
	1–19	3.49	< 20	1.90
	20–29	5.69	20–29	4.20
	40+	6.45		
ACS 50 State study	Non-smoker	1.00	Non-smoker	1.00
	All smokers	22.36	All smokers	10.80
			1–10	5.50
			11–19	11.20
			20	14.20
			21–30	20.40
			31+	22.00

Data from the new ACS Fifty State Study, however, show that the lung cancer risk among smokers has increased in more contemporary cohorts of both men and women (Table 2). Among current smokers, the lung cancer risk ratio has doubled among male smokers compared with results in the earlier ACS prospective study, increasing from approximately 11 to 22. The lung cancer risk for women has increased approximately four-fold from 3 to 12. This risk is almost identical to that observed among men in earlier studies initiated in the 1950s. Such changes indicate that among continuing smokers, the lung cancer risk has increased in more contemporary cohorts having a higher lifetime cigarette experience compared with cohorts enrolled in earlier studies. Ratios for most other cancer sites related to smoking have also increased in the period between studies, particularly for cancers of the respiratory system. The age-standardized lung cancer rate in non-smoking women, however, remained unchanged in the 26-year period covered by the two ACS studies (Figure 1). A similar pattern for men was reported in the 1989 Surgeon General's report (US Department of Health and Human Services, 1989).

Figure 1. Lung cancer death rates among smoking and non-smoking women over time adjusted to the 1970 US population

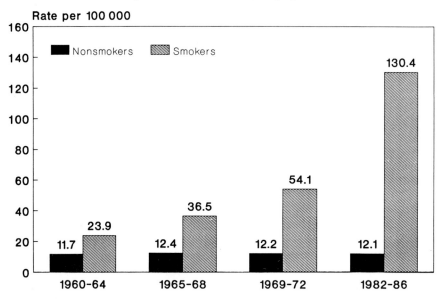

Source: Garfinkel and Stellmann (1988)

Cessation of smoking and reduced cancer risks

Data from the cohort mortality studies offer some insight into the benefits of stopping smoking compared with the risk of continuing to smoke. More recent data from the ACS Fifty State Study and the large case–control investigation of Lubin et al. (1984a) offer substantial insight into these risks among more recent groups of smokers and former smokers compared with studies initiated in the 1950s and 1960s.

Table 2 presents cancer mortality ratios from ACS Twenty-five State Study (initiated in 1959) and Fifty State Study (initiated in 1982) for both male and female current and former smokers. Data from the Twenty-five State Study is taken from the first six-year follow-up period rather than the 12-year period to make a more direct comparison with findings in the Fifty State Study. Current smokers from both studies are defined as 'cigarettes only' smokers. Ratios represent overall differences between continuing smokers and former smokers regardless of other factors such as length of time since last cigarette, number of cigarettes smoked per day, and duration of smoking.

Among males, the lung cancer risk for former smokers is less than half of that for current smokers. A similar finding is observed among women. For each cancer site, former smokers experience lower mortality compared with those classified as current smokers.

Table 2. Mortality ratios for males and females 35 years and older according to smoking status at time of enrolment

Cancer site	25-State study		50-State study	
	Current smoker	Former smoker	Current smoker	Former smoker
Males				
Lung	11.35	4.96	22.36	9.36
Oral	6.33	2.73	27.48	8.80
Oesophagus	3.62	1.28	7.60	5.83
Larynx	10.00	8.60	10.48	5.24
Bladder	2.90	1.75	2.86	1.90
Pancreas	2.34	1.30	2.14	1.12
Kidney	1.84	1.79	2.95	1.95
Females				
Lung	2.69	2.59[a]	11.94	4.96
Oral	1.96	1.89[a]	5.59	2.88
Oesophagus	1.94	2.15[a]	10.25	3.16
Larynx	3.81	3.10[a]	17.78	11.88
Bladder	2.87	2.31[a]	2.58	1.85
Pancreas	1.39	1.38[a]	2.33	1.78
Kidney	1.43	1.47[a]	1.41	1.16
Cervix	1.10	1.32	2.14	1.94

[a] Ratio is for current and former smokers combined.
Source: US Department of Health and Human Services (1989)

Time since smoking stopped and lung cancer

Data published over the past several decades have consistently shown that the longer one smokes cigarettes, the greater the lung cancer risk. Thus, any benefit from stopping smoking relating to reducing one's lung cancer risk is dependent to a large degree on the total duration of the behaviour before quitting. A number of prospective and case–control studies have examined this issue.

In all four earlier prospective studies (Table 3), lung cancer mortality risk was much lower among former male smokers than among continuing smokers, and this risk was strongly correlated with the length of time ex-smokers remained off cigarettes. In the US Veterans' (Rogot & Murray, 1980) and British Physicians' (Doll & Peto, 1976) studies, those former male smokers who had quit smoking and were able to stay off cigarettes for 15 years experienced lung cancer mortality risks between a half and a sixth of those observed in continuing smokers. However, even among former smokers who had quit for 15 or more years, lung cancer risk was two to four times higher than for lifelong non-smokers. Data from the Japanese study point in the same direction (Hirayama, 1974).

Thus, although giving up smoking appears to reduce one's risk of developing lung cancer, some elevation in risk may remain among former smokers for many years after quitting.

Table 3. Lung cancer mortality ratios in ex-cigarette smokers, by number of years stopped smoking

Study	Years stopped smoking	Mortality ratio	
British doctors	1–4	16.0	
	5–9	5.9	
	10–14	5.3	
	15+	2.0	
	Current smokers	14.0	
US veterans[a]	1–4	18.83	
	5–9	7.73	
	10–14	4.71	
	15–19	4.81	
	20+	2.10	
	Current smokers	11.28	
Japanese males	1–4	4.65	
	5–9	2.50	
	10+	1.35	
	Current smokers	3.76	
		Number of cigarettes smoked per day	
		1–19	20+
ACS 25 State (males 50–69)	< 1	7.20	29.13
	1–4	4.60	12.00
	5–9	1.00	7.20
	10+	0.40	1.06
	Current smokers	6.47	13.67

[a] Includes data only for ex-cigarette smokers who stopped for other than doctors' orders
Source: US Department of Health and Human Services (1982)

A consistent finding in the major prospective studies is that recent former smokers (those who had stopped smoking for less than five years) experienced much higher lung cancer mortality rates than other groups, including smokers who continued to smoke. In the US Veterans' study, for example, former smokers who had been off cigarettes for four years or less experienced a lung cancer mortality ratio of 18.8 compared with 11.3 for current smokers. A similar pattern is observed in other studies. This finding probably reflects that a certain proportion of smokers stopped smoking because they were diagnosed with lung cancer or exhibited symptoms of ill-health. With each additional time interval off cigarettes, however, the lung cancer mortality ratio between former smokers and never-smokers grows smaller. No data regarding lung cancer risk among former female smokers are available from these early studies.

A large case–control study by Lubin *et al.*, involved 7181 male and female lung cancer patients from five Western European countries admitted to hospitals between 1976 and 1980 (Lubin *et al.* 1984a). For each patient, two controls were selected and matched on a number of variables, including age at diagnosis, sex, centre, and category of hospital accommodation. Controls represented patients in whom diseases not related to tobacco use had been diagnosed. Patients were categorized according to smoking status and duration of smoking behaviour. For individuals who had stopped smoking, the length of time they had been off cigarettes was recorded. As smoking duration increased, the risk of developing lung cancer increased among those individuals who continued to smoke. Among ex-smokers, however, the risk declined for both men and women with the number of years off cigarettes, although among men the rate of decline was greater for those males who had smoked for a shorter length of time (Table 4). After ten years of not smoking, individuals who had smoked 19 years or less experienced a lung cancer risk similar to those who had never smoked. Risks remained elevated among both men and women who were classified as smoking for 20 years or more regardless of the time reported off cigarettes. Among men who had smoked the longest (greater than 50 years), even after stopping smoking for more than ten years, the lung cancer risk was not appreciably different from that of continuing smokers.

Table 4. Relative risk of developing lung cancer by time since stopping smoking and total duration of smoking habit

Time since stopping smoking (years)	Duration of smoking habit (years)			
	1–19	20–39	40–49	> 50
Men				
0	1.0[a]	2.2	2.8	3.0
1–4	1.1	2.1	3.3	3.8
5–9	0.4	1.5	2.2	2.8
> 10	0.3	1.0	1.6	2.7
Women				
0	1.0[b]	2.1	2.7	5.2
1–4	1.0	2.3	2.1	7.1
5–9	0.4	2.0	1.1	1.7
> 10	0.4	0.8	2.3	

[a] Baseline category. Risk for people who had never smoked relative to that for current smokers who had smoked for one to 19 years was 0.3.
[b] Baseline category. Risk for people who had never smoked relative to that for current smokers who had smoked for one to 19 years was 0.6.
Source: Lubin et al. (1984a)

These investigators also examined the impact of quitting smoking on lung cancer mortality by type of cigarette smoked, frequency and depth of inhalation, number of cigarettes smoked daily, and number of years off cigarettes (Table 5). The authors concluded: 'The results show that changes in patterns of cigarette smoking that lower exposure were associated with lower risks of developing lung cancer but, when compared with completely stopping smoking, the reductions in risk were small'.

Cessation of smoking and health status on lung cancer risk

Findings from existing studies make it appear that cessation of smoking behaviour does not result in any immediate decrease in lung cancer risk but rather suggest that an increase in risk may occur among recent ex-smokers. Because earlier studies did not take into account the importance of health status on reasons for stopping smoking, Garfinkel and Stellman (1988) analysed the health of 600 000 women at the time of their enrolment in the new ACS Fifty State

Table 5. Relative risk of developing lung cancer by number of years since stopping smoking, controlling for several other factors. All risks adjusted for duration of use in years

	Time since stopping smoking (years)							
	Men				Women			
	0	1–4	5–9	≥ 10	0	1–4	5–9	≥ 10
No. of cigarettes per day								
1–9	1.00[a]	1.47	0.90	0.67	1.00	1.55	1.16	0.66
10–19	1.00	1.31	0.88	0.61	1.00	1.04	0.62	0.20
20–29	1.00	1.08	0.75	0.51	1.00	0.95	0.64	0.33
> 30	1.00	0.86	0.80	0.40	1.00		0.24	0.42
Type of cigarette								
Filter	1.00	1.11	0.91	0.34	1.00	0.88	0.83	0.25
Mixed	1.00	0.98	0.69	0.53	1.00	0.73	0.61	0.27
Non-filter	1.00	1.12	0.64	0.33	1.00	2.16	0.65	0.30
Frequency of inhalation								
All the time	1.00	1.01	0.71	0.50	1.00	0.95	0.93	0.35
Most of the time	1.00	1.81	0.66	0.43	1.00	1.53	0.52	0.57
Part of the time	1.00	0.97	0.81	0.60	1.00	0.94	0.19	0.35
Rarely or never	1.00	1.13	0.69	0.39	1.00	1.49	1.02	0.29
Depth of inhalation								
Deeply	1.00	0.94	0.67	0.47	1.00	0.90	1.09	0.58
Moderately	1.00	1.22	0.73	0.43	1.00	1.01	1.68	0.47
Slightly or never	1.00	1.19	0.67	0.37	1.00	1.31	0.16	0.32

[a] Tests for differences in linear trends, $p < 0.05$
Source: Lubin et al. (1984a)

Study. Those who reported no history of heart disease, stroke or cancer were compared with all women in the study with respect to lung cancer risk among smokers and former smokers by the length of time since smoking and number of cigarettes smoked daily (Table 6). Similar data for men have not been reported from this study.

Among women who formerly smoked a packet or less of cigarettes daily and who reported no prior history of cancer or cardiovascular disease, lung cancer risk declined by nearly half within the first two years after quitting. This finding is in sharp contrast to findings among men in the earlier prospective studies when lung cancer risk was observed to increase in the first few years following cessation. Even heavier smokers (more than a packet of cigarettes daily) experienced a decline (from 27.5 to 24.0) in lung cancer risk within the first two years of cessation compared with continuing smokers if they reported no prior history of disease. Regardless of the amount smoked daily, with each additional time period off smoking, lung cancer mortality risk declined among ex-smokers relative to women who continued to smoke. However, even among lighter smokers (less than a packet daily), it appears that some degree of excess risk may remain even after more than 15 years off cigarettes. Women who consumed more than a packet daily experienced nearly a four-fold risk of dying from lung cancer compared with a never-smoker (Table 6). Although the mortality ratio does not fall completely to 1.0 even after 16 or more years of cessation, the number of deaths among women ex-smokers without a history of chronic disease was too small to predict reliably the number of years required to be off cigarettes for the risk to fall to that of a non-smoker.

Table 6. Expected (E) and observed (O) lung cancer deaths and standardized mortality ratios (SMR) among women who were former smokers, according to years since smoking stopped and number of cigarettes smoked at time of cessation

	No. of cigarettes per day		Never smoked regularly	Current smokers	Former smokers				
					Years since smoking stopped				
					0–2	3–5	6–10	11–15	16+
Women with no history of heart disease stroke or cancer	1–20	O	88	258	17	15	12	7	19
		E	88.0	17.8	1.9	1.9	3.1	3.7	13.3
		SMR	1.0	14.5	8.9	7.8	3.9	1.9	1.4
	21 or more	O		146	14	7	9	2	4
		E		5.3	0.6	0.5	0.7	0.8	1.1
		SMR		27.5	24.0	13.0	12.4	2.7	3.6
All women	1–20	O	174	335	52	33	20	21	41
		E	174.0	32.4	3.8	3.9	6.1	7.0	26.0
		SMR	1.0	10.3	13.6	8.4	3.3	3.0	1.6
	21 or more	O		195	39	23	17	6	9
		E		9.2	1.2	1.1	1.5	1.5	2.3
		SMR		21.2	32.4	20.3	11.4	4.1	4.0

Source: Garfinkel & Stellman (1988)

When the entire cohort of women is examined, including those with and without a prior history of disease, among both light and heavy smokers who quit, lung cancer mortality increased in the first two years after cessation, following the pattern observed for men in earlier prospective studies. Although the lung cancer risk declined as the interval off cigarettes increased, even after more than 15 years those women who smoked more than a packet of cigarettes daily were observed to have a risk for lung cancer four times higher than those who had never smoked. This study lends strong support to the conclusion that individuals who stop smoking because of existing disease or poor health status, possibly due to smoking, do not derive the same benefit from stopping as those who quit for other reasons.

Sixteen-year follow-up information from the US Veterans' study provides additional evidence to support this finding. Male smokers who gave 'doctor's orders' as their reason for stopping smoking experienced higher overall death rates than male veterans who stopped smoking for reasons 'other than doctor's orders'. Presumably those individuals who stopped smoking on advice of a doctor represented a group whose health status may have already been affected by smoking, whereas those individuals who gave up smoking for 'other than doctor's orders' may have chosen to stop out of concern for their future health.

Lung cancer and tar content of cigarettes smoked

Limited information is available in the scientific literature on the relationship between tar content of cigarettes smoked and a smoker's lung cancer risk (US Department of Health and Human Services, 1981). Low-tar-yield cigarettes today account for the largest share of the total USA cigarette market, with more than half of all cigarettes consumed in the USA in the low-tar category (US Department of Agriculture, 1987). Low-tar cigarettes are those with a tar content of 15 mg or less per cigarette. The Federal Trade Commission has estimated that the average sales-weighted tar content of cigarettes manufactured in the USA is about 12 to 13 mg tar (US Federal Trade Commission, 1985) compared with nearly 40 mg in the early 1950s (Wakeham, 1976). As late as 1972, only about 2% of the US cigarette market was classified as low tar although the tar content of cigarettes has been slowly declining since the introduction of filtered cigarettes in the mid-1950s (US Department of Health and Human Services, 1981). It can be concluded that the vast majority of smokers who currently smoke low-tar cigarettes changed to this type of product and started smoking with higher-tar-yield cigarettes.

Only one of the early major prospective studies examined lung cancer mortality by the tar and nicotine content of the cigarette smoked. In the ACS Twenty-five State Study, males who smoked low-tar and low-nicotine cigarettes experienced a lung cancer mortality rate 20% lower than males who continued to smoke higher-tar cigarettes. Female smokers of low-tar brands experienced nearly a 40% lower mortality rate than those smoking higher-tar brands. However, even smokers of low-tar cigarettes had lung cancer mortality rates that were significantly higher than those who had never smoked.

In the ACS Fifty State Study (Garfinkel & Stellman, 1988), the authors also examined the impact of low-yield cigarettes on the lung cancer risk of the 600 000 women in the cohort; data for men were not published. Slightly more than half (51.6%) of all women smoked cigarettes with less than 12 mg tar per cigarette, and only 3.6% smoked cigarettes with 20.2 mg tar or more. In this new analysis, a logistic regression model was developed whereby the tar yield of the cigarette smoked was a continuous variable together with categorical variables age (in five-year age groups), number of cigarettes consumed daily (grouped as none, 1–9, 10–19, 20, 21–30, and 31 or more) and degree of inhalation (none, slight, moderate, and deep). The outcome variable, died of lung cancer within four years, was placed at 1.0. For women cigarette smokers consuming cigarettes with tar yields of 5, 10, 15 and 20 mg, the lung cancer risks were 1.17, 1.36, 1.59 and 1.85 times greater than non-smokers, respectively. These investigators concluded that 'doubling the tar yield would be equivalent to an increased risk of about 40%, independently of amount smoked or depth of inhalation'.

Several retrospective studies in the United States (Wynder & Stellman, 1979), Great Britain (Rimington, 1981) and Austria (Kunze & Vutuc, 1980) have found that smokers who consumed filter-tipped cigarettes experience lower lung cancer mortality than smokers of non-filtered cigarettes. More recently, Lubin et al. (1984b), in a large case–control study carried out in several Western European countries, found that lifelong non-filter smokers were at nearly twice the risk of lung cancer compared to lifelong filter smokers even after controlling for the duration of cigarette use and number of cigarettes smoked per day. Presumably filtered cigarettes trap a certain proportion of the tar and overall are lower in tar content than non-filtered cigarettes. Today, more than 90% of all cigarettes manufactured in the USA are filter-tipped (US Department of Health and Human Services, 1981; US Department of Health, Education and Welfare, 1979).

Considerable controversy exists in the scientific community about the significance of reduced tar content in light of revelations that some cigarette brands have been purposely engineered to test low on smoking machines (Bell, 1983; Wald, 1985). A considerable scientific research base has now proved that the manner in which machines test tar does not accurately reflect how smokers actually smoke this new generation of low-yield cigarettes (US Department of Health and Human Services, 1981). Thus, although the overall reduction in tar yield may somewhat reduce the risk of lung cancer and possibly other smoking-related cancers, it is prudent to assume that proportionate reduction in tar is probably not correlated with a proportionate reduction in cancer risk. From both a public health and an individual perspective, the only definite way one can reduce lung cancer risk is not to take up smoking or, if one does smoke, to give up smoking.

Histological changes in the lung attributed to cessation or reduction in cigarette tar content

The classic autopsy studies of Auerbach et al. (1957, 1962) provide evidence that stopping smoking results in a lower risk of developing histological changes in the lung compared with continued smoking. Auerbach found that the frequency and intensity of change correlate with the degree of smoking exposure as measured by the number of cigarettes smoked per day. Changes in the bronchial epithelium where all cells were classified as atypical were present in 4.3% of whole lung sections in males who consumed one to two packets of cigarettes daily, in 11.4% of those smoking more than two packets daily, and in more than 14% among those diagnosed as having died of lung cancer. Such changes were not observed among men classified as non-smokers and were observed in only 0.2% of former smokers. For the total series, only 4% of former smokers were observed to have any histological change compared with 17 and 38%, respectively, of those smoking less than two and more than two packets daily and 1% among males who had never smoked regularly.

In another study, Auerbach *et al.* (1979) compared the degree of histological change of men who died in the period 1955 to 1960 with those who died during 1970 to 1977. These investigators found that smokers in the latter group exhibited less advanced histological changes than the former group. The authors attributed this finding to the reduced tar and nicotine yield of cigarettes smoked by this group compared with the average yield of those smoked by the earlier group.

References

American Cancer Society (1989) *Cancer Facts and Figures - 1989*. Atlanta, American Cancer Society

Auerbach, O., Gere, J.B., Forman, J.B., Petrick, T.G., Smolin, H.J., Muehsam, G.E., Kassonry, D.Y. & Stout, A.P. (1957) Changes in the bronchial epithelium in relation to smoking and cancer of the lung. *New Engl. J. Med.*, 256, 97-104

Auerbach, O., Stout, A.P., Hammond, E.G. & Garfinkel, L. (1962) Changes in bronchial epithelium in relation to sex, age, residence, smoking and pneumonia. *New Engl. J. Med.*, 267, 111-119

Auerbach, O., Hammond, E.C., Garfinkel, L. & Parks, V.R. (1979) Changes in bronchial epithelium in relation to cigarette smoking, 1955-1960 vs 1970-1977. *New Engl. J. Med.*, 300, 381-386

Bell, J. (1983) Government to review tar-table tests. *New Scientist*, 100, 480

Doll, R. & Peto, R. (1976) Mortality in relation to smoking: 20 years' observations on male British doctors. *Br. Med. J.*, 2, 1525-1536

Doll, R. & Peto, R. (1981) The causes of cancer: Quantitative estimates of avoidable risks of cancer in the United States today. *J. Natl Cancer Inst.*, 66, 1191-1308

Doll, R., Gray, R., Hafner, B. & Peto, R. (1980) Mortality in relation to smoking: 22 years' observations on female British doctors. *Br. Med. J.*, 280, 967-971

Fortune Magazine (1935) The Fortune Survey. III. Cigarettes. *Fortune Magazine*, 12, 111-116

Garfinkel, L. & Stellman, S.D. (1988) Smoking and lung cancer in women: findings in a prospective study. *Cancer Res.*, 48, 6951-6955

Haenszel, W., Shimkin, M.B. & Miller, H.P. (1956) Tobacco smoking patterns in the United States. *Public Health Service Monograph No. 45* (Public Health Service Publication No. 463), Bethesda, MD, US Department of Health, Education and Welfare

Hammond, E.C. (1966) Smoking in relation to the death rates of one million men and women. In: Haenszel, W., ed., *Epidemiological Approaches to the Study of Cancer and Other Diseases*. (National Cancer Institute Monograph No. 19), Bethesda, MD, US Public Health Service, pp. 127-204

Hammond, E.C., Garfinkel, L., Seidman, H. & Lew, E.A. (1976) 'Tar' and nicotine content of cigarette smoke in relation to death rates. *Environmental Res.*, 12, 263-274

Hammond, E.C., Garfinkel, L., Seidman, H. & Lew, E.A. (1977) Some recent findings concerning cigarette smoking. In: Hiatt, H.H., Watson, J.D. & Winsten, J.A., eds, *Origins of Human Cancer. Book A: Incidence of Cancer in Humans* (Cold Spring Harbor Conference on Cell Proliferation, Vol. 4) Cold Spring Harbor, NY, CSH Press, pp. 101-112

Harris, J.E. (1983) Cigarette smoking among successive birth cohorts of men and women in the United States during 1900-80. *J. Natl Cancer Inst.*, 71, 473-479

Hirayama, T. (1974) Prospective studies on cancer epidemiology based on census population in Japan. In: Bucalossi, P., Veronesi, U. & Cascinelli, N., eds, *Cancer Epidemiology, Environmental Factors*, Vol. 3 (Proceedings of the 11th International Cancer Congress, Florence, Italy, October 20-26, 1974), pp. 26-35

Hirayama, T. (1977) Changing patterns of cancer in Japan with special reference to the decrease in stomach cancer mortality. In: Hiatt, H.H., Watson, J.D. & Winsten, J.A., eds, *Origins of Human Cancer. Book A: Incidence of Cancer in Humans* (Cold Spring Harbor Conference on Cell Proliferation, Vol. 4), Cold Spring Harbor, NY, CSH Press, pp. 55-75

IARC (1986) *IARC Monographs on the Evaluation of the Carcinogenic Risk of Chemicals to Humans*. Vol. 38, *Tobacco Smoking*, Lyon, International Agency for Research on Cancer

Kahn, H.A. (1966) The Dorn study on smoking and mortality among US Veterans: Report on 8 1/2 years of observation. In: Haenszel, W., ed., *Epidemiological Approaches to the Study of Cancer and Other Diseases* (National Cancer Institute Monograph No. 19), Bethesda, MD, US Public Health Service, pp. 1-125

Kunze, M. & Vutuc, C. (1980) Threshold of tar exposure: Analysis of smoking history of male lung cancer cases and controls. In: Gori, G.B. & Bock, F.G., eds, *Banbury Report 3 - A Safe Cigarette?*, Cold Spring Harbor, NY, CSH Press, pp. 29-36

Lubin, J.H., Blot, W.J., Berrino, F., Lamant, F., Gillis, C.R., Kunze, M., Schmahl, D. & Visco, G. (1984a) Modifying risk of developing lung cancer by changing habits of cigarette smoking. *Br. Med. J.*, 288, 1953-1956

Lubin, J.H., Blot, W.J., Berrino, F., Gillis, C.R., Kunze, M., Schmahl, D. & Visco, G. (1984b) Patterns of lung cancer risk according to type of cigarette smoked. *Int. J. Cancer*, 33, 569-576

McGinnis, J.M., Shopland, D. & Brown, C. (1987) Tobacco and health: Trends in smoking and smokeless tobacco consumption in the United States. In: Breslow, L., Fielding, J.E. & Lave, L.B., eds, *Annual Review of Public Health*, Vol. 8, Palo Alto, Annual Reviews, pp. 441-467

National Center for Health Statistics (1986) Health United States - 1986. And Prevention Profile. *DHHS Publication No. (PHS) 87-1232*, Bethesda, MD, US Department of Health and Human Services, Public Health Service

Rimington, J. (1981) The effect of filters on the incidence of lung cancer in cigarette smokers. *Environmental Res.*, 24, 162-166

Rogot, E. & Murray, J.L. (1980) Smoking and causes of death among US Veterans: 16 years of observation. *Public Health Reports*, 95, 213-222

Shopland, D.R. (1986) Smoking. In: Bernstein, E., ed., *Medical and Health Annual*, Chicago, IL, Encyclopedia Britannica Inc., pp. 420-424

Shopland, D.R. (1987) Smoking and health: A twenty year reflection. In: Rosenberg, M.J., ed., *Smoking and Reproductive Health*, Littleton, MA, PSG, pp. 4-8

US Department of Agriculture (1987) Tobacco situation and outlook report. Washington, DC, US Department of Agriculture, Economic Research Service

US Department of Health, Education and Welfare (1979) *Smoking and Health. A Report of the Surgeon General* (DHEW Publication No. (PHS) 79-50066), US Department of Health Education and Welfare, Public Health Service, Office of Smoking and Health

US Department of Health and Human Services (1980) *The Health Consequences of Smoking for Women. A Report of the Surgeon General*, Rockville, MD, US Department of Health and Human Services, Public Health Service, Office on Smoking and Health

US Department of Health and Human Services (1981) *The Health Consequences of Smoking: The Changing Cigarette. A Report of the Surgeon General* (DHHS (PHS) 81-50156), Washington, DC, US Department of Health and Human Services, Public Health Service, Office on Smoking and Health

US Department of Health and Human Services (1982) *The Health Consequences of Smoking: Cancer. A Report of the Surgeon General* (DHHS Publication No. (PHS) 82-50179), Rockville, MD, US Department of Health and Human Services, Public Health Service, Office on Smoking and Health

US Department of Health and Human Services (1989) *Reducing the Health Consequences of Smoking: 25 Years of Progress. A Report of the Surgeon General, 1989* (DHHS Publication No. (CDC) 89-8411), Rockville, MD, US Department of Health and Human Services, Public Health Service, Office on Smoking and Health, Centers for Disease Control

US Federal Trade Commission (1985) Report to Congress on the Cigarette Labelling and Advertising Act of 1969

Wakeham, H. (1976) Sales weighted average tar and nicotine deliveries of US cigarettes from 1957 to present. In: Wynder, E.L. & Hecht, S., eds, *Lung Cancer* (UICC Technical Report Series, Vol. 25), Geneva, International Union Against Cancer, pp. 151-152

Wald, N. (Chairman) Participants of the fourth Scarborough Conference on Preventive Medicine. (1985) Is there a future for low tar-yield cigarettes? *Lancet*, ii, 1111-1114

Wynder, E.L. & Stellman, S.D. (1979) Impact of long-term filter cigarette usage on lung and larynx cancer risk. A case-control study. *J. Natl Cancer Inst.*, 62, 471-477

CHANGES IN DIET AND CHANGES IN CANCER RISK: OBSERVATIONAL STUDIES

D.M. Parkin and M.P. Coleman

*Unit of Descriptive Epidemiology,
International Agency for Research on Cancer,
Lyon, France*

The proportion of cancer mortality which may be attributable to dietary factors, at least in the USA, has been estimated at 35%, though with a wide margin of uncertainty from 10% to 70% (Doll & Peto, 1981). The equivalent proportion of cancer incidence due to diet has been estimated at about 40% and 60% for males and females, respectively (Wynder & Gori, 1977), while Higginson and Muir (1979) gave estimates of 30% and 63% for Birmingham, UK, and 18% and 58% for Bombay, India, attributed more generally to lifestyle factors, including diet. Dietary influences on cancer risk were estimated by these authors to be at least as important as tobacco exposure, yet the role of diet in the causation of cancer has been difficult to study, partly because 'diet' encompasses such a wide variety of foods and dietary habits, but also because the means by which these habits can be measured are cumbersome and difficult to apply to large numbers of individuals. These difficulties are reflected in the uncertainty surrounding the overall risk of cancer that is attributable to what we eat.

Despite the difficulties in interpreting the evidence on diet and cancer, several agencies have issued dietary guidelines intended to reduce cancer risk. A committee of the US National Research Council (NRC) conducted a comprehensive review of diet and nutrition in the etiology of cancer. It concluded that cancers at most major sites were influenced by diet, but that the evidence was still inadequate either to quantify its contribution to cancer etiology or to determine the reduction in risk that might be achieved by dietary change. The NRC nevertheless issued interim dietary guidelines judged to be 'consistent with good nutritional practices and likely to reduce the risk of cancer' (NRC, 1982). The guidelines stressed reduction of total fat intake, daily consumption of fruit, vegetables and whole-grain cereal products, avoidance of foods preserved by salt-curing, salt-pickling or smoking, and consumption of alcoholic beverages (if at all) only in moderation. Patterson and Block (1988) have shown, however, that among North Americans aged 19–74 who were surveyed with a single 24-hour dietary recall in 1976–80 for the National Health and Nutrition Examination Survey (NHANES II), before the NRC guidelines were issued, less than a quarter reported consuming any foods in the three food groups considered as protective by the NRC in 1982 – namely, cruciferous vegetables (18%), fruits and vegetables rich in vitamin A (21%), and high-fibre cereals or whole-grain bread (16%). The NRC guidelines did not include avoidance of obesity, although five subsequent sets of dietary guidelines on cancer, issued in the USA, Japan and Europe, included the maintenance of appropriate body weight (Palmer, 1986). All these guidelines were consistent with the NRC recommendations; some also included limitation of sodium chloride intake.

Additional NRC policy recommendations were aimed at reducing contamination of foods with carcinogens, monitoring the food supply for conformity with permissible levels of contaminants, testing new additives for carcinogenicity, and removing mutagens from foods, if feasible.

The NRC stressed the interim nature of its guidelines, and that they should be reconsidered every few years in the light of new evidence. There is still considerable controversy over the issue, however, with opinions ranging from a belief in avoidance of any policy without better evidence, on the one hand, to the view that more definitive and stringent recommendations on the reduction of dietary fat are needed, on the other (Palmer, 1986). It seems clear that if recommendations to change the way we eat are to be widely adopted, the evidence on which they are based will first need to be widely accepted. This requires evaluation of the evidence.

In this chapter we restrict our attention mainly to the limited evidence from observational (non-intervention) studies that changes in diet may affect cancer risk. We also review the evidence linking changes in cancer risk to alcohol consumption patterns, and to those of a commonly occurring carcinogenic contaminant of the diet, aflatoxins.

Almost all studies of cancer risk in relation to changes in consumption of certain foods or drinks are population-level correlations. Such studies provide only weak evidence for causation or prevention, but they may be useful in gauging the extent to which risk might be reduced by any plausible changes in the dietary habits of a whole population. The relevant criterion to assess any benefit to public health from dietary change must be the extent to which an observed reduction in cancer risk can be attributed to the dietary change. Evidence from epidemiological studies concerning a particular dietary habit or component as a risk factor for a particular cancer is of course relevant, but it does not follow that the proportion of cancer risk 'attributable' to diet can necessarily be prevented. Wahrendorf (1987) has shown, for example, by applying a simple model to the distribution of risk from several case–control studies of cancers of the stomach, colon and rectum, that the proportion of these three cancers which might be prevented by any plausible shift in the population distribution of dietary habits could be as low as 15–20%, whereas Doll and Peto (1981) 'guestimated' that up to 90% of deaths from these cancers might be avoidable by practicable dietary means. The wide disparity between these estimates of the attributable and – more practically – the preventable proportions of cancer, suggests that there is room for further consideration of the evidence.

Cancer risk in individuals in relation to changes in diet

We are aware of three studies which provide evidence concerning dietary change in individuals and subsequent cancer risk. Two of these concern selected persons who decided, mostly in adulthood, to adopt religious lifestyles incorporating dietary restrictions. Phillips (1975) has described cancer mortality in a cohort of 35 000 Seventh-Day Adventists in California, whose church advocates a lacto-ovo-vegetarian diet rich in fruit, vegetables and whole-grain cereals, and avoidance of tobacco, alcohol and coffee. Some adult members are ex-smokers and about half continue to eat some meat, though much less often than comparable non-Adventists. Total cancer mortality at seven years' follow-up was significantly lower than in California as a whole (SMR 53 in males, 67 in females), and although this was largely due to very low mortality from cancers related to tobacco and alcohol, mortality from cancer of the colon, stomach and pancreas was also 20 to 50% lower than in the general population. There was, however, no relationship between meat intake and mortality from these cancers within the Adventist population (Phillips & Snowdon, 1983). Examination of cancer risk by age at entry to the church showed that age-adjusted mortality at age 35 and over from colorectal and breast cancer was not significantly different among members who joined at age 35 years or over from those who joined before the age of 18, and mortality from these cancers was somewhat higher in 'lifetime' members than in those who had been members for 12 years or less. A more recent analysis also found that the risk of fatal breast cancer was not related to age at first exposure to a vegetarian lifestyle, nor to its duration (Mills et al., 1988). Kinlen (1982) described cancer mortality in a retrospective cohort study covering a 68-year period up to 1978 in two groups of over 1000 nuns living in closed communities in Britain, one of which avoided meat completely, while the other allowed some meat in the diet. There were no significant overall differences from the general population in mortality from colorectal and breast cancer, and no evidence that duration of exposure to a meat-free or low-meat diet had any effect on mortality from these cancers. These studies do not provide any evidence that reducing meat and fat consumption in adult life will reduce cancer mortality, but remain consistent with the possibility that dietary exposures in childhood may be important in determining cancer risk.

Hirayama (1982) investigated the influence of a change in the frequency of consumption of green-yellow vegetables on subsequent stomach cancer mortality, by means of a re-survey of a subset (3%) of a large cohort of Japanese adults, six years after study entry. For the group who had increased their frequency of intake, stomach cancer mortality was almost the same as in frequent consumers, and significantly lower than in those who continued not consuming these vegetables. The finding was confined to males, and implies that, whatever the beneficial components, their effect on risk is rather rapid.

Cancer risk in populations in relation to changes in diet

The evidence from population-level studies on the effects of dietary change on cancer risk derives from studies relating time trends in cancer incidence or mortality to dietary changes, and from studies of migrant populations.

Time trends in cancer risk in relation to diet in populations

Observation of changes over time in the diet of a population and in its risk of various cancers presents certain difficulties. In principle, cancer incidence data would be preferable to mortality rates for the examination of trends in risk, but cancer registration is a relatively recent development in many countries, and it takes some time to achieve reasonably reliable registration. Trends in recorded incidence may thus be partly due to improvements over time in the completeness of cancer registration, as well as to changes in classification of malignancy for sites such as the cervix or bladder. For most countries, cancer mortality data are the only index of risk available over a sufficiently long period for the examination of time trends. Doll and Peto (1981) preferred mortality to incidence data as being more complete and more 'trustworthy', but trends in mortality may not always be a suitable index of the underlying trends in cancer incidence: trends in mortality data may reflect trends in incidence less well as survival improves, and this will differ between cancer sites. Changes in the nature or quality of death certification may also affect the interpretation of trends in cancer risk, and an example of this problem is discussed below.

The defects of population-level correlation studies in etiological research are well known (see for example Breslow & Enstrom, 1974). The major one is, of course, the so-called 'ecological fallacy' – the fact that associations between disease and exposures in a population may not reflect those observed in the individual members of the population. The number of different exposures for which data are available for the populations studied may be relatively few, so that it is difficult to allow for the effects of confounding. Armstrong and Mann (1985) have discussed the limitations of dietary data in ecological studies, in particular, their crudity: usually gross national statistics on agricultural production or food available for consumption (Robertson & Kevany, 1982), without age- or sex-specific figures. As with cancer mortality at different sites, the quality of the available data – in the sense that it reflects precisely what we want to know – is likely to differ from one food or nutrient to another, thus affecting observed correlations with disease. The quality and detail of dietary data may also vary with economic development, which may produce spurious associations with dietary variables, such as meat intake, which are linked to economic development. Some studies have correlated current dietary patterns with indices of current cancer risk, although it is clearly more appropriate to relate cancer risk to dietary patterns prevalent a number of years earlier. Doing so, however, means that an additional arbitrary element – a fixed latency period (generally 10–30 years) – is introduced into the correlations.

In fact, only a few studies have attempted to correlate trends in cancer incidence with trends in dietary factors, and they did not all incorporate a time-lag. The kinds of data available on food consumption patterns from routine sources will not always be the most relevant for studies of diet and cancer. For example, cereal, fruit and vegetable intakes are only imprecise surrogates for the vegetable fibre components considered most relevant to large bowel function and cancer risk (Bingham et al., 1979), and the strength of any underlying trends observed with these surrogate variables will probably be reduced.

The strong cross-sectional geographical correlation between aggregate measures of fat or meat consumption and national mortality rates for breast cancer was first observed by Lea (1966) and later confirmed by many workers. He found similar correlations between countries for fat intake and cancers of the ovary and rectum, and for leukaemia at ages 55 and over. Correlations of breast cancer mortality and fat consumption have also been noted between regions within countries. Correlations have been reported between meat or fat intake in 32 countries and mortality from cancers of the colon and rectum, ovary, uterus, prostate and testis (Armstrong & Doll, 1975).

Hems (1978) examined correlations between breast cancer mortality, diet and family size in 41 countries, finding large positive correlations with breast cancer for total fat, particularly of animal origin, and a weaker association with family size. For 20 countries for which sufficient data were available, time trends in breast cancer mortality for the period 1950–67 were correlated with time trends in *per caput* consumption of dietary items for the period 1948–65, each trend estimated by linear regression. The trends in breast cancer mortality were significantly and positively correlated with trends in estimated consumption of animal protein ($r = 0.55$) and fat ($r = 0.51$).

McMichael *et al.* (1979) investigated time trends in mortality rates from colon and rectal cancer in relation to patterns of food and alcohol consumption in the United States, United Kingdom, Australia and New Zealand. In all four countries, mortality from colon cancer (and less markedly rectal cancer) fell during the 1940s and 1950s – a change involving both sexes and all age groups. The declines were most marked in England and Wales, and suggested possible beneficial effects of the low-fat, low-sugar, high-fibre diet of war-time Britain. However, subsequent changes in the UK, and the time trends in the other countries cannot convincingly be related to consumption patterns of meat, fat or fibre. A somewhat more plausible relationship can be found with patterns of beer consumption in the different countries (see below).

Time trends in breast cancer risk and in consumption of fat, meat and dairy products have been documented for the Nordic countries between 1934 and 1980. Breast cancer incidence rates increased in all five countries up to 1980, with a steady rise in incidence in successive ten-year cohorts born between 1882 and 1932. In Finland, both fat consumption and breast cancer incidence were the lowest of the Nordic countries at the beginning of the available data periods, and both increased more than in the other countries in the period up to 1980, but the differentials and time trends in breast cancer risk between the Nordic countries 'do not correspond exactly' with the differentials and trends in dietary factors (Hakulinen *et al.*, 1986).

Trends in age-standardized breast cancer mortality and diet in England and Wales have been studied by several authors. Ingram (1981) examined the period 1928–77, and reported a sharp decrease in breast cancer mortality at the beginning of the second world war, at a time when there were equally sudden changes in national dietary patterns, particularly major falls in consumption of fat (18%), meat (30%) and sugar (49%), and an increase in cereal and vegetable consumption. Reversal of the dietary changes by about 1954 was followed by a rise in breast cancer mortality until 1970. The correlation coefficient with breast cancer mortality was maximal for fat consumption (0.87) with a lag-time of 12 years, but coefficients for sugar and meat were very similar, with slightly different lag-times. All three variables were closely inter-correlated. These results were interpreted as evidence for an effect of war-time dietary changes on breast cancer risk. Hems (1980) had noted, however, that the decline in breast cancer mortality during the period 1911 to 1975 began in the 1930s, before the introduction of rationing in 1940; Key et al. (1987) later showed that much of the sharp decline in recorded breast cancer mortality can be ascribed to changes introduced in 1940 in the rules for selecting the underlying cause of death, as a result of which cancer was no longer given priority if it occurred as one of several causes on the certificate. After adjustment of the trends for this artefact, the decrease in mortality between 1930 and 1950 is shown to begin before the war, and to be largely confined to the 50–69 years age group, suggesting a cohort effect. The overall upward trends in fat intake and breast cancer mortality in England and Wales during this century remain compatible with a causal link, but the evidence from dietary changes specifically related to the second world war is insufficient to conclude that deliberate reduction in dietary fat would reduce incidence or mortality from breast cancer.

The effects of famine on age at menarche and other risk factors for breast cancer are being studied among women who experienced the severe famine which occurred in the western part of the Netherlands in 1944–45. At a national level, the breast cancer mortality rates of the cohort born around 1930 are no different from those of adjacent cohorts (Figure 1), but among women screened for breast cancer in the region affected by the famine, the breast cancer detection rate does appear to show a cohort effect (van Noord et al., 1988). Women born during 1917–31, for a proportion of whom the menarche might have been delayed by the famine, have a lower breast cancer detection rate in each of three age groups than those in a similar cohort who would have passed menarche before the famine, even though any effect of famine on breast cancer risk would be diluted by the fact that not all the women in the screening area were living there during the famine. The effects of this natural experiment are being studied further, and the availability of individual data on exposure to famine will offer some advantages over the more usual correlation studies in detecting any effect of the severe dietary shortages on subsequent cancer risk; however, the famine occurred during major social upheaval caused by war, and the changes in quantity and type of food eaten may not be directly relevant to voluntary changes in diet.

Data on trends in fat, meat and fibre consumption in relation to breast and colon cancer mortality have been presented by Armstrong and Mann (1985) for Australia and the USA. The patterns appear confusing: breast cancer mortality in the USA has shown little change since the 1930s, though fat consumption has been increasing steadily, while a post-war decline in Australia has reversed since 1960, although fat intake has increased only since 1970.

In summary, although there are apparently rather clear population-level correlations between the consumption of different dietary items and the risk of

Figure 1. Female breast cancer mortality, Netherlands, birth cohort 1910-1945

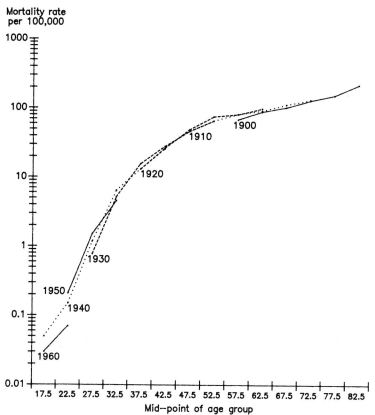

different cancers, there are few opportunities to extend such analyses to include a temporal dimension. The lack of clear-cut results is perhaps not surprising given the inherent weaknesses of population-level studies and the technical problems discussed above. In addition, trends in food or nutrient consumption patterns are generally unidirectional (increases in meat and fat and decreases in fibre) so that, if there were a resultant change in cancer risk, this would be manifest as a rising or falling trend, susceptible to many other possible explanations.

Migrant studies

Studies of migrant populations provide, in theory at least, the opportunity to observe the results of a natural experiment, where a population of individuals has elected to live in a new environment. The risk of cancer in these populations can be compared with that in the country of origin (genetically similar population in a different environment) or in the host country (genetically different, but similar environment). The studies can be refined by looking at cancer risk in the migrant generation itself in comparison to their offspring (born in the new country), or, if the necessary data are available, in relation to the time since migration (or duration in the new environment) of the migrants. These refinements permit inferences to be made about the importance of the duration of environmental exposure (or, equivalently, the age of the individual at the time of environmental change).

The 'natural experiment' provided by the study of human migrant populations differs from the ideal experimental model in two major respects. Firstly, the population who migrate are not a random sample of the population of the country of origin. Migrants may be more (or less) healthy than the average, of higher or lower social status, or even composed of particular ethnic or religious subgroups (as in the case of Jewish migrants to Israel, for example), and may therefore have cancer experiences quite different from the general source population. In addition, the region of origin of migrants is usually unknown, and this may be important where there is considerable regional variation in risk of certain cancers, such as for nasopharyngeal cancer in China. Secondly, the extent to which migrants are exposed to the 'environment' of the host country is usually unknown; the external environment (the air, or the water, or levels of background radiation) may be very similar, but in other respects – particularly cultural practices such as diet, many of the habits of the country of origin may be retained, even after several generations.

Changes in incidence and mortality rates in migrants for those cancers presumed to be related primarily to dietary factors – notably stomach, colo-rectal and breast cancer – have been of particular interest to epidemiologists. Unfortunately, in the great majority of studies, it is *assumed* that migrants have changed their diet towards the pattern in the host country. Very few studies have actually documented what type of dietary change occurs on migration, or the differences in diet between country of origin, migrant groups, and the host country. The main reason is that routinely collected data on the diet of individuals are quite rare, and surveys are unlikely to ask for country of birth of the individuals questioned. Special ad-hoc studies of diet in migrant populations are not frequent either, although the data from control subjects in some case–control studies, notably in Hawaii (see below), can be used to make inferences about dietary habits in different ethnic groups.

A review of migrant studies will not be attempted here – interested readers should consult Kmet (1970) and Haenszel (1982).

The principal groups of migrants that have been studied are populations of Japanese origin in the United States (Haenszel & Kurihara, 1968) and Hawaii (Tominaga, 1985), populations of Chinese origin in the United States (King *et al.*, 1985), populations of Latin-American origin in the United States (Mack *et al.*, 1985), European migrants to the United States (Nasca *et al.*, 1981) and migrants to Britain (Marmot *et al.*, 1984), Australia (Armstrong *et al.*, 1983), and Israel (Halevi *et al.*, 1971; Steinitz *et al.*, 1989).

In the review below, we will consider only those studies where the extent of dietary change in migrants has been related to changes in cancer risk.

Stomach cancer

All published studies to date involve migrants from countries with relatively high risk of stomach cancer, moving to areas of low risk. In general, the pattern of change has been a relatively slow decline in risk towards that of the host country in first-generation migrants, with the risk in subsequent generations approaching that of the host country more closely.

Perhaps the most comprehensive data concern the populations of Japanese origin in Hawaii, who have emigrated from a high-risk country to an area of low risk. Figure 2 shows rates for Japanese and whites in Hawaii, by place of birth, and for comparison rates for US whites and for Japan (Kolonel et al., 1981). Hawaiian-born Japanese have rates which are lower than those for first-generation migrants (born in Japan), while the rates for Hawaiian-born whites are higher than those for whites born elsewhere. Comparison of these data with interview data on diet (Table 1) suggests that second-generation Japanese eat less pickled vegetables and dried salted fish than Japanese migrants born in Japan, whereas the whites born in Hawaii seem to eat these foods more than whites born elsewhere. Both food items have been associated with stomach cancer risk in case–control studies (Haenszel et al., 1972; Bjelke, 1974). Similar findings were noted for rice consumption (and total carbohydrate intake) – again suggested as etiologically important in various studies (Segi et al., 1957; Graham et al., 1967). However, fruit consumption was high in Japanese migrants born in Japan and whites born in Hawaii, as was vitamin C intake (in females only); in theory, both of these dietary components should have been protective against stomach cancer risk.

Figure 2. Age-adjusted stomach cancer incidence in Hawaii Japanese and Caucasians by place of birth, 1973–1977

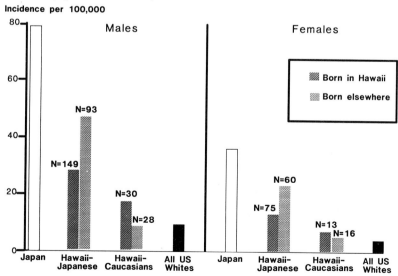

Sources: Hawaiians, Kolonel et al. (1981); US whites (1973–1977), Young et al. (1981); Japan (1975), Hanai et al. (1984)

Table 1. Mean dietary intakes in relation to stomach cancer incidence by race, sex and place of birth[a]

	Born in Hawaii	Stomach cancer incidence[b]	Dried salt fish	Pickled vegetables	Rice
Japanese male	Yes	112	0.1	2.8	7.4
	No	156	0.2	3.1	8.7
Japanese female	Yes	48	0.1	2.8	7.2
	No	54	0.3	4.1	7.5
Caucasian male	Yes	61	0.1	0.4	5.3
	No	31	0.0	0.3	2.3
Caucasian female	Yes	23	0.0	0.3	4.3
	No	16	0.0	0.1	1.8

[a] Intakes expressed as average number of times consumed per week (data taken from Kolonel et al., 1981).
[b] Age-standardized rate per 100 000 adjusted to world population aged 45 and over.

In Australia, migrant populations have rates intermediate between those of the country of origin and the (lower) rates in Australia — for most populations, the decline in risk is related to the duration of stay in Australia (Table 2). McMichael et al. (1980) have examined these data in relation to consumption patterns of various foodstuffs and alcohol in the countries of origin and in Australia some 10 years earlier (in the 1950s). All countries of origin had a much higher intake of complex, unrefined carbohydrates, such as cereals, potatoes and nuts, than Australia, and all except Greece had lower fruit consumption. Consumption of spirits in Poland and Yugoslavia was much higher than in Australia, and wine consumption was higher in Greece and Italy (the UK and Irish patterns were rather similar to those found in Australia). More recent data (McMichael & Giles, 1988) are available from a large national survey in 1983 comparing the diets of Australians with migrants from southern Europe (Italy, Yugoslavia, Greece); they suggest that migrants maintain some elements of a Mediterranean diet (higher consumption of bread, pasta, leafy vegetables and red wine) more than the Australian-born do. It is impossible to be certain, however, about how much change has occurred in migrants' diets, but a comparison of the recent study and earlier 1950s surveys (Table 3) suggests that the migrants may move towards Australian patterns more for some dietary items such as meat, milk, cheese and alcoholic beverages, than for others (cereals and fruit).

Colon and rectal cancers

In general, most studies have involved migrant populations moving from low-risk to high-risk areas, such as Chinese and Japanese to Hawaii or the USA, or Southern Europeans to Australia. In general terms, the risk of colon cancer approximates fairly rapidly to that in the new country, and, in the case of Japanese in the United States, incidence rates of both colon and rectal cancer in those born in the USA now exceed the rates of US whites (Shimizu et al., 1987). The same is true of Japanese males living in Hawaii (Table 4).

Table 2. Risk of stomach cancer: country of origin, migrants, Australia

Country of origin	Standardized rate ratio (Australia = 1.0)			
	Country of origin	Migrants < 16 yrs	Migrants > 16 yrs	Australia
England	1.55	1.47	1.24	1.0
Scotland	1.67	1.84	1.46	1.0
Ireland	1.88	1.77	1.21	1.0
Poland	3.38	1.69	1.71	1.0
Yugoslavia	2.30	2.22	1.23	1.0
Greece	1.34	1.35	1.15	1.0
Italy	2.19	1.43	1.49	1.0

Figures are ratios of age-sex-standardized death rates: country of origin *versus* Australia, for ages 35-64, 1970-71; migrants to Australia *versus* Australian-born, for ages 30+, 1962-76 (from McMichael *et al.*, 1980)

Table 3. Consumption patterns of foodstuffs and alcoholic drinks in Southern Europe, relative to Australia, during the 1950s (McMichael *et al.*, 1980) and in migrants from Southern Europe to Australia, relative to Australian born, 1983 (McMichael & Giles, 1988).

Dietary item	Southern Europe[a], 1950s (reference: Australia = 1.0)	S. European migrants, 1983 (reference: Australia born = 1.0)
Cereals	1.5	
Bread, pasta		2.2
Potatoes	1.0	0.6
Meat	0.2	
Beef, veal, poultry, sausages		1.0
Vegetables	1.4	
Leafy greens, tomatoes, carrots, pumpkins		0.9
Fruits	1.0	
Citrus fruits		1.6
Milk and cheese	0.4	0.7
Beer	0.04	0.5
Wine	8.6	1.4

Figures are ratios of estimated daily consumption per person.
[a] Southern Europe data are calculated as weighted averages of data from Italy, Greece and Yugoslavia, using the proportions of migrants from each in Australia during 1962-1971 (according to Armstrong *et al.*, 1983)

Table 4. Incidence of cancer in Japan and Hawaii (1973-1977) (age standardized incidence (world standard) per 100 000)

Tumour primary site		Japan[1] (Miyagi)	Hawaii Japanese[2]		Hawaii[1] whites
			Issei	Nisei	
Stomach	M	88.0	46.9	28.5	12.2
	F	42.0	22.9	13.2	5.0
Colon	M	8.3	29.2	28.0	25.3
	F	7.3	19.4	18.8	17.8
Rectum	M	9.2	24.0	17.6	14.1
	F	6.5	9.2	7.9	8.9
Breast	F	17.5	35.9	57.2	85.6

[1] From Waterhouse et al. (1982)
[2] From Kolonel et al. (1980)

Several studies of Japanese migrants to Hawaii allow the nature and extent of dietary change to be evaluated. Comparison of Japanese men in Hiroshima and Hawaii (Kagan et al., 1974) showed that men in Hawaii consume more animal protein, cholesterol and total fat, and less carbohydrate, vegetable protein and alcohol than men in Japan. Interview data from the Japan–Hawaii cancer study suggested that, for meat intake, Hawaiian Japanese were more similar to Hawaiian whites than to rural Japanese, while milk intake was intermediate between the two populations (Stemmermann et al., 1979). At the same time, it was noted that Hawaiian Japanese and whites had similar daily stool weight (62% of that in rural Japanese); however, bowel transit time was similar in the two Japanese populations, and some 50–60% that of whites in Hawaii.

Migrants to Australia from low-risk countries show quite marked increases in the risk of colon cancer with increasing duration of stay, and changes are more marked for males than females, particularly for southern European migrants (McMichael et al., 1980). Migrant rectal cancer mortality shows a similar pattern of convergence towards Australian-born rates (this involves a decline in risk for males and females born in England). As described earlier, the extent of dietary change in migrants is not clear, but there do appear to have been increases in consumption of meat and dairy produce in southern European migrants, as well as a large increase in beer consumption, which has been suggested as an etiological agent in rectal cancer. The larger changes in mortality rates in males than in females would be in keeping with probably larger changes in the diets of men, who, in full-time employment, would be more exposed to indigenous canteen cuisine.

Breast cancer

Most migrant populations have moved from areas of low incidence to areas of higher incidence for breast cancer. In general, the risk of breast cancer appears to change relatively quickly towards that of the host country, although transition to the high rates of US whites may take several generations in Japanese and Chinese migrants (Dunn, 1977). At present, the incidence rates of breast cancer in US Japanese populations remain about one-half to two-thirds those of whites in the same areas, but are approximately double the rates observed in Japan (Muir et al., 1987). Japanese born in Hawaii (*Nisei*) have higher incidence rates than first-generation migrants (*Issei*), although rates remain lower than those of the white population (Table 4). In a cross-sectional survey comparing the differences in incidence with the diet of the two generations of migrants, Kolonel et al. (1980) found the increased risk of breast cancer (almost 60%) inconsistent with the small decreases (5%) in total fat intake in women over 45 years of age between the two generations. However, later data from the same source (Hankin et al., 1983) using a different method of age-adjustment suggest significant increases in the intake of total fat (+5.5%) and animal fat (+4%) between the two generations of females.

The relative importance of dietary change and change in other risk factors (parity, age at first birth, age at menarche, age at menopause) is not clear from migrant studies. However, for Italian-born migrants to Australia, mortality rates in women over 40 reach Australian levels within 5–20 years of arrival, and it is unlikely that a change in age at first pregnancy could account for this (McMichael & Giles, 1988).

Alcohol

The evidence for the role of alcohol in human carcinogenesis has been reviewed recently (IARC, 1988). Many observational studies, of both case–control and cohort design, have demonstrated increased risks of cancers of the oral cavity, pharynx, oesophagus and larynx; alcoholic beverages with high alcohol concentration may carry a higher risk than those of low concentration. Although alcohol acts in a multiplicative manner with smoking, there is evidence of an effect of alcohol in non-smokers. Associations between alcohol intake and increased risk of cancer of the stomach, rectum, liver, pancreas and breast remain less certain.

We are unaware of any intervention studies in which the alcohol intake of individuals or populations has been reduced in order to study benefits to health. None of the many observational studies showing an association between alcohol intake and cancer risk (referred to above) appear to have included a category of ex-drinkers so that, in analogy to ex-smokers, the possible decline in risk following cessation of alcohol consumption could be assessed.

The only data on the effects of declining alcohol consumption on risk of cancer are from observations of human populations in whom there have been 'natural experiments' in terms of reduced alcohol consumption. Tuyns and Audigier (1976) analysed male mortality from cancer in France in relation to trends in national alcohol consumption. A marked decline in consumption occurred in relation to the second world war during 1941–46, which was accompanied by a fall in the number of deaths due to alcoholism and liver cirrhosis. They noted that the cohorts of males born in 1902–16, who were aged 25–40 at the time of this reduced alcohol consumption, showed a decline in the risk of laryngeal cancer and a halt in the increasing rates of oesophageal cancer. No such changes were observed for cancers of the lung or pancreas.

McMichael (1978, 1979) has made similar observations for Britain and Australia. Time trends in mortality from laryngeal and oesophageal cancer are similar, and unlike those of lung cancer. Alcohol consumption in Australia fell to a minimum during the Depression in the early 1930s, before it began rising progressively. Generations of males and females who were young adults at that time subsequently show the lowest age-specific mortality at each successive age for laryngeal cancer, but there has been a tendency for mortality rates in all age groups to rise since the 1950s, possibly due to the participation of each generation in the post-war upsurge in alcohol consumption.

Chilvers *et al.* (1979) compared time trends in alcohol intake in England and Wales (1900–70) with mortality from oesophageal cancer (1911–75). They found a very close correlation between standardized cohort mortality ratios for the age range 35–74 years and the *per caput* consumption of alcohol, particularly spirits and beer, between 25 and 29 years of age (Figure 3).

Finally, McMichael *et al.* (1979) have drawn attention to the similarity in the trends of beer consumption since 1900 in four countries (USA, UK, Australia and New Zealand), and trends in colon and rectal cancer mortality some 20–30 years later. Early in the century, beer consumption in the UK was much higher than in the other countries, but it declined until 1930, and has increased only slightly since then. Colon cancer mortality in the UK was very high before 1940–50, particularly in men, but rates then fell, most markedly in males, with a reversed sex ratio by the 1970s. In the other three countries, beer consumption rose in 1935 to 1945, and mortality from colon cancer increased after 1950; however, whereas in the USA beer consumption then stabilized (with little change in colon cancer mortality since 1960), in Australia and New Zealand consumption has continued to increase, as have mortality rates from colon cancer and, since 1960, from rectal cancer.

Data such as these are suggestive that a decline in alcohol-related cancer incidence and mortality might be expected to follow decreased alcohol consumption at the population level. Further evidence is available from the comparisons of first-generation (*Issei*) and second-generation (*Nisei*) Japanese living in Hawaii (Kolonel *et al.*, 1980). There has been a decline in the percentage of regular alcohol drinkers between generations (20% in men and 24% in females), along with corresponding changes in the incidence of oesophageal and rectal cancers.

Aflatoxin

Aflatoxin is considered to be a hepatic carcinogen, mainly on the basis of correlation studies in Africa and Asia (IARC, 1987). Aflatoxin is found in many dietary items, particularly grains and peanuts, which have been contaminated with

Figure 3. Trends in oesophageal cancer mortality and alcohol consumption in England and Wales

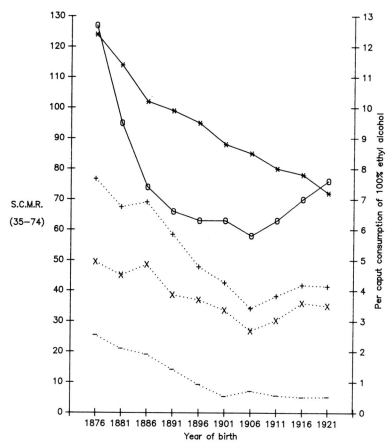

———, oesophageal cancer mortality; O, males; ✶, females

........, alcohol consumption; +, all alcoholic beverages; X, beer; -, spirits

the fungus *Aspergillus flavus*. As a contaminant of diet, therefore, there is scope for preventive action in reducing the degree of contamination. Measures such as improvements in harvesting, drying and storage of crops, and avoidance of obviously mouldy food are appropriate in the areas most at risk, the developing world, where the population lives on the food it grows itself (see Linsell, 1979). The effectiveness of intervention measures is at present unknown. van Rensburg et al. (1985) have observed a decline in hepatocellular carcinoma (HCC) in Inhambane province, Mozambique, between 1968 and 1973. If real, this could hardly be related to change in chronic carriage of hepatitis-B surface antigen, the major known risk factor for HCC; they speculate that it is due to decreased aflatoxin consumption following publicity about the dangers of mouldy food, and increased selectivity in preparation of foodstuffs, or to declining consumption of peanuts. The same authors cite the survey of Purves (1973) that shows a reduced risk of liver cancer in gold-miners six months to one year after commencing employment, indicating that reduced aflatoxin intake (the miners' commercial food supplies are under statutory control) may have a rapid effect in reducing risk. Bradshaw *et al.* (1982) also recorded a fall in liver cancer incidence from 68.6 to 40.8 per 100 000 man-years (crude rates) between 1964–71 and 1972–79 among black gold-miners from Mozambique, a much greater proportional decrease than in miners from nine other territories working in South Africa, and showed that the decline was occurring throughout this period. They also speculate about the possible influence of a reduction in hepatic carcinogens or promoters in food as a result of improved storage.

If these speculations are true, it should be relatively easy to investigate the effects of intervention on liver cancer incidence in a high-risk area. The IARC–UNEP project designed to assess the effects of the Rural Development Areas (RDA) programme in Swaziland has shown that it is possible to reduce the aflatoxin content of diet; the reduction in aflatoxin contamination of foodstuffs (except peanuts) between two survey periods (1972–73 and 1982–83) was most marked in areas with the highest level of intervention (the programme included improved methods of harvesting and storage) (Bosch & Muñoz, 1989). However, no assessment of the possible impact on incidence of liver cancer has been made.

Acknowledgements

We wish to thank Professor A.J. McMichael and Professor L.N. Kolonel for helpful comments and suggestions during the preparation of this chapter.

References

Armstrong, B.K. & Doll, R. (1975) Environmental factors and cancer incidence and mortality in different countries. *Int. J. Cancer, 15*, 617-631

Armstrong, B.K. & Mann, J.I. (1985) Diet. In: Vessey, M.P. & Gray, M., eds, *Cancer Risks and Prevention*, Oxford, Oxford University Press, pp. 68-98

Armstrong, B.K., Woodings, T.L., Stenhouse, N.S. & McCall, M.G. (1983) *Mortality from Cancer in Migrants to Australia – 1962 to 1971*, Nedlands, WA, University of Western Australia

Bingham, S., Williams, D.R.R., Cole, T.J. & James, W.P.T. (1979) Dietary fibre and regional large-bowel cancer mortality in Britain. *Br. J. Cancer, 40*, 456-463

Bjelke, E. (1974) Epidemiologic studies of cancer of the stomach, colon and rectum, with special emphasis on the role of diet. *Scand. J. Gastroenterol., 9*, (Suppl. 31), 1-253

Bosch, F.X. & Muñoz, N. (1989) Epidemiology of hepatocellular carcinoma. In: Bannasch, P., Keppler, D. & Weber, G., eds, *Liver Cell Carcinoma* (Falk Symposium No. 51), Dordrecht, Kluwer, pp. 3-14

Bradshaw, E., McGlashan, N.D., Fitzgerald, D. & Harington, J.S. (1982) Analyses of cancer incidence in black gold miners from Southern Africa (1964-79). *Br. J. Cancer, 46*, 737-748

Breslow, N.E. & Enstrom, J.E. (1974) Geographic correlations between cancer mortality rates and alcohol-tobacco consumption in the United States. *J. Natl Cancer Inst., 53*, 631-639

Chilvers, C., Fraser, P. & Beral, V. (1979) Alcohol and oesophageal cancer: an assessment of the evidence from routinely collected data. *J. Epidemiol. Commun. Health, 33*, 127-133

Doll, R. & Peto, R. (1981) The causes of cancer. *J. Natl Cancer Inst., 66*, 1191-1308

Dunn, J.E. (1977) Breast cancer among American Japanese in the San Francisco Bay Area. *Natl Cancer Inst. Monogr., 47*, 157-160

Graham, S., Lilienfeld, A.M. & Tidings, J.E. (1967) Dietary and purgation factors in the epidemiology of gastric cancer. *Cancer*, 20, 2224-2234

Haenszel, W. (1982) Migrant studies. In: Schottenfeld, D. & Fraumeni, J.F., eds, *Cancer Epidemiology and Prevention*, Philadelphia, W.B. Saunders

Haenszel, W. & Kurihara, M. (1968) Studies of Japanese migrants 1. Mortality from cancer and other diseases among Japanese in the United States. *J. Natl Cancer Inst.*, 40, 43-68

Haenszel, W., Kurihara, M., Segi, M. & Lee, R.K.C. (1972) Stomach cancer among Japanese in Hawaii. *J. Natl Cancer Inst.*, 49, 969-988

Hakulinen, T., Andersen, AA., Malker, B., Pukkala, E., Schou, G. & Tulinius, H. (1986) Trends in cancer incidence in the Nordic countries: a collaborative study of the five Nordic cancer registries. *Acta Pathol. Microbiol. Immunol. Scand.*, Sect. A, Suppl. 288, Vol. 94

Halevi, H.S., Dreyfuss, F., Peritz, E. & Schmelz, U.O. (1971) Cancer mortality and immigration to Israel 1950-67. *Isr. J. Med. Sci.*, 7, 1386-1404

Hanai, A., Kitamura H., Fukuma, S. & Fujimoto, I. (1984) *Cancer Incidence in Japan 1975-1979*, Osaka, Osaka Cancer Registry

Hankin, J.H., Kolonel, L.N., Yano, K., Heilbrun, L. & Nomura, A.M.Y. (1983) Epidemiology of diet-related diseases in the Japanese migrant population of Hawaii. *Proc. Nutr. Soc. Aust.*, 8, 22-40

Hems, G. (1978) The contributions of diet and childbearing to breast-cancer rates. *Br. J. Cancer*, 37, 974-982

Hems, G. (1980) Associations between breast-cancer mortality rates, child-bearing and diet in the United Kingdom. *Br. J. Cancer*, 41, 429-437

Higginson, J. & Muir, C.S. (1979) Environmental carcinogenesis: misconceptions and limitations to cancer control. *J. Natl Cancer Inst.*, 63, 1291-1298

Hirayama, T. (1982) Does daily intake of green-yellow vegetables reduce the risk of cancer in man? An example of the application of epidemiological methods to the identification of individuals at low risk. In: Bartsch, H. & Armstrong, B., eds, *Host Factors in Human Carcinogenesis* (IARC Scientific Publications No. 39), Lyon, International Agency for Research on Cancer, pp. 531-540

IARC (1987) Aflatoxins. In: *IARC Monographs on the Evaluation of Carcinogenic Risks to Humans*, Suppl. 7, *Overall Evaluations of Carcinogenicity: An Updating of IARC Monographs Volumes 1-42*, Lyon, International Agency for Research on Cancer, pp. 83-87

IARC (1988) *IARC Monographs on the Evaluation of Carcinogenic Risks to Humans*, Vol. 44, *Alcohol Drinking*, Lyon, International Agency for Research on Cancer

Ingram, D.M. (1981) Trends in diet and breast cancer mortality in England and Wales 1928-1977. *Nutr. Cancer*, 3, 75-80

Kagan, A., Harris, B.R., Winkelstein, W., Johnson, K.G., Kato, H., Syme, S.L., Rhoods, G.G., Gay, M.L., Nichaman, M.Z., Hamilton, H.B. & Tillotson, J. (1974) Epidemiologic studies of coronary heart disease and stroke in Japanese men living in Japan, Hawaii and California: demographic, physical, dietary and biochemical characteristics. *J. Chron. Dis.*, 27, 345-364

Key, T.J.A., Darby, S.C. & Pike, M.C. (1987) Trends in breast cancer mortality and diet in England and Wales from 1911 to 1980. *Nutr. Cancer*, 10, 1-9

King, H., Li, J.Y., Locke, F.B., Pollack, E.S. & Tu, J.-T. (1985) Patterns of site-specific displacement in cancer mortality among migrants: the Chinese in the United States. *Am. J. Publ. Health*, 75, 237-242

Kinlen, L.J. (1982) Meat and fat consumption and cancer mortality: a study of strict religious orders in Britain. *Lancet*, i, 946-949

Kmet, J. (1970) The role of migrant populations in studies of environmental effects. *J. Chron. Dis.*, 23, 293-304

Kolonel, L.N., Hinds, M.W. & Hankin, J.H. (1980) Cancer patterns among migrant and native-born Japanese in Hawaii in relation to smoking, drinking, and dietary habits. In: Gelboin, H.V., MacMahon, B., Matsushima, T., Sugimura, T., Takayama, S. & Takebe, H., eds, *Genetic and Environmental Factors in Experimental and Human Cancer*, Tokyo, Japan Scientific Societies Press, pp. 327-340

Kolonel, L.N., Nomura, A.M.Y., Hirohata, T., Hankin, J.H. & Hinds, M.W. (1981) Association of diet and place of birth with stomach cancer incidence in Hawaii Japanese and Caucasians. *Am. J. Clin. Nutr.*, *34*, 2478-2485

Lea, A.J. (1966) Dietary factors associated with death rates from certain neoplasms in man. *Lancet*, *ii*, 332-333

Linsell, C.A. (1979) Decisions on the control of a dietary carcinogen – aflatoxin. In: Davis, W. & Rosenfeld, C., eds, *Carcinogenic Risks. Strategies for Intervention* (IARC Scientific Publications No. 25), Lyon, International Agency for Research on Cancer, pp. 111-122

Mack, T.M., Walker, A., Mack, W. & Bernstein, L. (1985) Cancer in Hispanics in Los Angeles county. *Natl Cancer Inst. Monogr.*, *69*, 99-104

Marmot, M.G., Adelstein, A.M. & Bulusu, L., eds (1984) *Immigrant mortality in England & Wales 1970-78* (Studies on Medical and Population Subjects No. 47), London, Office of Population Censuses and Surveys

McMichael, A.J. (1978) Increases in laryngeal cancer in Britain and Australia in relation to alcohol and tobacco consumption trends. *Lancet*, *i*, 1244-1247

McMichael, A.J. (1979) Laryngeal cancer and alcohol consumption in Australia. *Med. J. Aust.*, *1*, 131-134

McMichael, A.J & Giles, G.G. (1988) Cancer in migrants to Australia: extending the descriptive epidemiological data. *Cancer Res.*, *48*, 751-756

McMichael, A.J., Potter, J.D. & Hetzel, B.S. (1979) Time trends in colo-rectal cancer mortality in relation to food and alcohol consumption: United States, United Kingdom, Australia and New Zealand. *Int. J. Epidemiol.*, *8*, 295-303

McMichael, A.J., McCall, M.G., Hartshorne, J.M. & Woodings, T.L. (1980) Patterns of gastrointestinal cancer in European migrants to Australia: the role of dietary change. *Int. J. Cancer*, *5*, 431-437

Mills, P.K., Annegers, J.F. & Phillips, R.L. (1988) Animal product consumption and subsequent fatal breast cancer risk among Seventh-day Adventists. *Am. J. Epidemiol.*, *127*, 440-453

Muir, C., Waterhouse, J., Mack, T., Powell, J. & Whelan, S., eds (1987) *Cancer Incidence in Five Continents*, Volume V (IARC Scientific Publications No. 88), Lyon, International Agency for Research on Cancer

Nasca, P.C., Greenwald, P., Burnett, W.S., Chorost, S. & Schmidt, W. (1981) Cancer among the foreign-born in New York State. *Cancer*, *48*, 2323-2328

NRC, Committee on Diet, Nutrition and Cancer, Assembly of Life Sciences (1982) *Diet, Nutrition and Cancer*, Washington DC, National Academy Press

Palmer, S. (1986) Dietary considerations for risk reduction. *Cancer*, *58*, 1949-1953

Patterson, B.H. & Block, G. (1988) Food choices and the cancer guidelines. *Am. J. Public Health*, *78*, 282-286

Phillips, R.L. (1975) Role of life-style and dietary habits in risk of cancer among Seventh-Day Adventists. *Cancer Res.*, *35*, 3513-3522

Phillips, R.L. & Snowdon, D.A. (1983) Association of meat and coffee use with cancers of the large bowel, breast, and prostate among Seventh-Day Adventists: preliminary results. *Cancer Res.*, *43*, 2403s-2408s

Purves, L.R. (1973) Primary liver cancer in man as a possible short duration seasonal cancer. *S. Afr. J. Sci.*, *69*, 173-178

Robertson, J.A. & Kevany, J.J. (1982) Food consumption patterns within the EEC. *Irish J. Med. Sci.*, *151*, 366-373

Segi, M., Fukushima, I., Fujisaku, S., Kurihara, M., Saito, S., Asano, K. & Kamoi, M. (1957) An epidemiological study on cancer in Japan. *Gann*, 48 (Suppl.), 1-63

Shimizu, H., Mack, T.M., Ross, R.K. & Henderson, B.E. (1987) Cancer of the gastrointestinal tract among Japanese and white immigrants in Los Angeles county. *J. Natl Cancer Inst.*, *78*, 223-228

Steinitz, R., Parkin, D.M., Young, J.A., Bieber, C.A. & Katz, L., eds (1989) *Cancer in Jewish Migrants to Israel* (IARC Scientific Publications No. 98), Lyon, International Agency for Research on Cancer

Stemmermann, G.N., Mandel, M. & Mower, H.F. (1979) Colon cancer: its precursors and companions in Hawaii Japanese. *Natl Cancer Inst. Monogr.*, *53*, 175-179

Tominaga, S. (1985) Cancer incidence in Japanese in Japan, Hawaii and the western United States. *Natl Cancer Inst. Monogr.*, *69*, 83-92

Tuyns, A.J. & Audigier, J.C. (1976) Double wave cohort increase for oesophageal and laryngeal cancer in France in relation to reduced alcohol consumption during the Second World War. *Digestion*, *14*, 197-208

van Noord, P.A.H., Collette, H.J.A., de Waard, F. & Rombach, J.J. (1988) Caloric restriction in early life: effects on breast cancer risk factors? In: Riboli, E. & Saracci, R., eds, *Diet, Hormones and Cancer: Methodological Issues for Prospective Studies* (IARC Technical Report No. 4), Lyon, International Agency for Research on Cancer, pp. 112-119

van Rensburg, S.J., Cook-Mozaffari, P., van Schalkwyk, D.J., van der Watt, J.J., Vincent, T.J. & Purchase, I.F. (1985) Hepatocellular carcinoma and dietary aflatoxin in Mozambique and Transkei. *Br. J. Cancer*, *51*, 713-726

Wahrendorf, J. (1987) An estimate of the proportion of colo-rectal and stomach cancers which might be prevented by certain changes in dietary habits. *Int. J. Cancer*, *40*, 625-628

Waterhouse, J., Muir, C., Shanmugaratnam, K. & Powell, J., eds (1982) *Cancer Incidence in Five Continents*, Volume IV (IARC Scientific Publications No. 42), Lyon, International Agency for Research on Cancer

Wynder, E.L. & Gori, G.B. (1977) Contribution of the environment to cancer incidence: an epidemiologic exercise. *J. Natl Cancer Inst.*, *58*, 825-832

Young, J.L., Percy, C.L. & Asire, A.J., eds (1981) *Surveillance, Epidemiology, and End Results: Incidence and Mortality Data, 1973-77* (National Cancer Institute Monograph No. 57), Washington, DC, US Government Printing Office

Evaluating Effectiveness of Primary Prevention of Cancer
Ed. M. Hakama, V. Beral, J.W. Cullen & D.M. Parkin
Lyon, International Agency for Research on Cancer
© IARC, 1990

PREVENTION OF CANCER: REVIEW OF THE EVIDENCE FROM INTERVENTION TRIALS

M.P. Coleman[1] and M. Law[2]

[1]*International Agency for Research on Cancer,
Lyon, France*
[2]*St Bartholomew's Medical College, London, UK*

Perhaps the most persuasive evidence for the efficacy of a measure designed to prevent cancer would be obtained from a well executed randomized, controlled trial in which cancer risk was reduced by the intervention. Such evidence would be all the more valuable if it concerned an exposure for which the etiological evidence was still incomplete, and the intervention could be readily adopted by large sections of the population to prevent a common cancer. A randomized, controlled trial of anti-smoking advice to current smokers has been reported (Rose *et al.*, 1982; see below), but the evidence that tobacco smoke is a powerful carcinogen is already strong, and in most populations only about a third or a half of adults are current smokers. Dietary factors are likely to be implicated in at least a third and perhaps as much as two thirds of human cancer (see chapter by Parkin & Coleman in this volume), so evidence that a simple dietary intervention could reduce cancer risk would be particularly useful.

In this chapter we examine the results of controlled trials involving a change in diet or lifestyle factors which have been demonstrated to be related to the risk of cancer. All of the trials were aimed at the prevention of coronary heart disease (CHD). Trials of drugs are not discussed, and issues related to chemoprevention trials (the use of micronutrients as inhibitors of carcinogenesis) are dealt with by Buring and Hennekens in this volume. It is not our purpose to confirm or refute evidence of causality derived from longitudinal observational studies, but rather to review whether health education or other interventions are a useful tool in cancer control. For this to be so, good compliance leading to long-lasting changes in relevant lifestyles is required, in addition to an etiological role for the factors concerned.

Smoking

The only single-factor, randomized, controlled trial of anti-smoking advice to current smokers was reported by Rose *et al.* (1982); anti-smoking advice in other trials has always been part of a multifactorial intervention. In this trial, 1445 male smokers aged 40–59 years were selected as being at the highest risk of cardiorespiratory disease, on the basis of a multivariate risk factor assessment, from among 16 016 men screened in 1968–70 for the Whitehall Study of London civil servants. They were randomized to an intervention group (714 men) and a 'normal care' or control group (731 men), and randomization produced groups that were well matched for age (mean: 53 years), weight and smoking habit (mean: 19 cigarettes per day). Men in the intervention group were recalled to receive individual advice about smoking and health on an average of five occasions over the first 12 months; men in the control group were not contacted, and seldom received anti-smoking advice from their general practitioner. The men were followed up for ten years, during which postal questionnaire data on smoking were obtained on three occasions, after one year, three years and nine years. Deaths up to 1979 and cancer registrations (1971–79) were obtained by flagging the study population in the National Health Service Central Register. Over the ten-year period, men in the intervention group reduced their cigarette consumption by about 60%, compared to 28% among the controls.

Ten-year mortality from all causes was similar in the two groups. The ten-year actuarial mortality for CHD was 18% lower in the intervention group (49 versus 62 deaths), and 23% lower for lung cancers (18 versus 24 deaths), although neither difference was significant at the 5% level (Table 1). Cancers at sites other than the lung, however, were twice as frequent in the intervention group as among controls (41 versus 19 cases). Although this excess was statistically significant at the 1% level, the authors argued that it was still likely to be due to chance. First, relative to mortality in England and Wales as a whole, there was a deficit of other cancers in the control group (12 deaths observed versus 20.8 expected, SMR 58), as well as an excess in the intervention group (28 versus 19.9 deaths, SMR 141); the differences between observed and expected deaths contributed about equally to the overall difference between intervention and control groups (see Table 1). Second, the lung cancer risk in both study groups was lower in those who stopped smoking than in those who continued to smoke, whereas the excess risk of other cancers between intervention and control groups was apparent both among those who stopped smoking and among those who did not. Further, the excess of cancers in the intervention group was not concentrated among any particular group of sites, whether related to smoking or not. These points were taken as evidence that the excess of cancers other than lung cancer in the intervention group was not related to cessation of smoking. Some further light is thrown on these results by recent data from the MRFIT trial (see chapter by Friedewald *et al.* in this volume).

Trials of diet and cancer prevention

No randomized trial of a dietary intervention designed primarily to reduce cancer risk in individuals has yet been carried out. The first full-scale, primary prevention trial of diet and cancer with incident cancers as the primary endpoint will probably be the Women's Health Trial, designed to test the hypothesis that a low-fat diet will reduce the incidence of breast cancer among women at high risk (Clifford *et al.*, 1986; Greenwald *et al.*, 1987). Up to 30 000 participants will be recruited from among motivated women aged 45–69 with an interest in early detection of breast cancer (contacted, for example, via screening projects and health maintenance organizations). Eligible women will have either two or more

Table 1. Mortality and cancer incidence after ten years in the Whitehall trial of anti-smoking advice (Rose et al., 1982)

	Intervention			Normal care			Total	
	Cases and deaths	Deaths[a]		Cases and deaths	Deaths[a]		Deaths[a]	
		O	E		O	E	O	E
Cancer								
Lung	22	18	17.2	25	24	18.1	42	35.3
Other sites	41	28	19.9	19	12	20.8	40	40.7
Smoking-related[b]		(11)	(5.6)		(2)	(5.9)	(13)	(11.5)
Remainder		(17)	(14.3)		(10)	(14.9)	(27)	(29.2)
Cardiovascular								
CHD		49			62		111	94
Other cardiovascular		13			12		25	
Stroke		(7)			(5)		(12)	
Remainder		(6)			(7)		(13)	
Other causes		15			18		33	
Chronic bronchitis		(3)			(4)		(7)	
Remainder		(12)			(14)		(26)	
Total		123			128		251	257

[a] Observed (O) in the trial and expected (E) on the basis of age-specific mortality rates for England and Wales, 1974.
[b] Cancers of the upper respiratory tract (ICD8 140, 143-9), oesophagus (150), pancreas (157), urinary tract (188-9).

female relatives with breast cancer, or at least two characteristics from among the following:

(a) benign breast disease (2 biopsies);

(b) breast cancer in one female relative;

(c) nulliparous or first-term birth aged 30 or more;

(d) atypical epithelial hyperplasia (1 biopsy).

The intervention will take the form of intensive counselling intended to reduce dietary fat intake by half, from the typical level of 40% of calories to about 20% of calories. Follow-up will last for at least eight to ten years. A feasibility study of 303 women has been reported (Thompson, 1987) in which 184 women randomized to the intervention reduced their fat intake from 39.1% to 21% of calories and maintained this reduction over 12 months, while fat intake for control women was virtually unchanged (38.9% and 38.1%), thus confirming at least that the intervention is practical. At the time of writing it remains unclear whether the main study will be funded (estimates exceed US $10 million).

Trials of coronary heart disease prevention

Several trials have been carried out to assess the feasibility of primary or secondary prevention of coronary heart disease (CHD). The dietary intervention in these studies mostly took the form of advice to substitute polyunsaturated fat for saturated fat in the diet, or simply to reduce total fat intake; additional advice given in some studies concerned smoking, obesity and control of blood pressure. None of the trials was designed to study the effect of the intervention on cancer incidence or mortality and cancer endpoints may therefore not have been recorded with the same degree of care as CHD events (see following chapter). It is still of interest to examine the cancer results from these trials, however. There is some evidence in experimental animals to suggest that an increase in polyunsaturated fat intake might increase cancer risk (Gammal et al., 1967), but while Broitman et al. (1977) found that diets high in unsaturated fats increased the yield of colonic tumours in rats after induction by dimethylhydrazine, Reddy et al. (1976) found similar increases with both saturated and unsaturated fats. In these animal experiments, dietary fats appeared to act as promoters; if similar mechanisms were to operate in human carcinogenesis, the effect of reduction of dietary fat intake on cancer incidence might occur within a relatively short period of time. Direct experimental evidence in humans of any effect of dietary fat manipulation on the subsequent risk of cancer is thus of considerable importance. Any evidence that an intervention which reduced CHD risk might have an adverse effect on cancer risk would be equally important, and CHD prevention studies are worth examining for these reasons. Three trials of CHD prevention (excluding drug trials) have reported unexpected adverse effects on mortality, two of them for cancer mortality: both these studies are discussed below.

Some of the CHD trials and other intervention studies which have been done are reported elsewhere in this book, and they are therefore mentioned only briefly in this chapter: these are the WHO collaborative trial in the multifactorial prevention of CHD (WHO European Collaborative Group), the North Karelia study (Hakulinen et al.) and the Multiple Risk Factor Intervention Trial (MRFIT; Friedewald et al.).

In the Oslo Diet-Heart study (Leren, 1966, 1970), 412 men aged 30-64 were recruited after their first myocardial infarction during the period 1956-58 and randomized to receive repeated dietary advice to adopt a low-cholesterol diet over a five-year period: mean serum cholesterol reductions over this period were 17.6% and 3.7% in the intervention and control groups, respectively. After 11 years, about half the men in each group had died. There was a significant advantage for the intervention group in fatal myocardial infarction (32 versus 57 deaths), but no significant difference in the number of cancers (7 versus 5 deaths), all of which were confirmed at autopsy.

The MRC trial of soya-bean oil (Research Committee, 1968) was designed to test whether a cholesterol-lowering diet would reduce the occurrence rates of second heart attack in men who survived a first one. The trial was carefully executed, and it achieved a 22% reduction in serum cholesterol in the intervention group (versus 6% in the controls), but there was no difference in the rates of re-infarction or CHD mortality. The trial was small. Over the six-year follow-up period, death from carcinoma occurred in one of 199 men in the intervention group and in 6 of 194 controls.

In the Veterans Administration trial (Dayton et al., 1968), 846 men aged 54–88 years (mean: 65 years) living in an institution were randomized either to conventional diet (40% of calories as fat) or to a cholesterol-lowering diet in which total fat was still about 39% of calories, but animal fat was largely substituted by vegetable oils. The two diets were served in separate cafeterias, but were made to resemble each other closely, and the trial was conducted 'double-blind' (Dayton et al., 1969). Adherence to the regime, assessed as the percentage attendance record in the two cafeterias, was not quite so good among experimental subjects (49%) as among controls (56%), but the average decrease in serum cholesterol was still 12.7% greater for the experimental group than for controls. Average body weight remained similar in both groups. After eight years there was a significant reduction among the experimental group in mortality from heart disease, sudden death and stroke combined (48 versus 70 deaths); this was essentially confined to men aged less than 65 years at entry. At the end of the period, however, there was a small excess of cancer deaths (7 versus 2 all cancers; 5 versus 1 lung cancer) in the intervention group, causing the authors to raise the question of possible adverse effects of the intervention.

In a later report (Pearce & Dayton, 1971), both incident and fatal cancers were studied. During the diet phase, there were 57 carcinomas (31 deaths) in the experimental group, and 35 carcinomas (17 deaths) in the control group. The cancer mortality of the two groups diverged steadily throughout the eight-year dietary intervention and this difference was not explained by differences in smoking habits: controls smoked slightly but significantly more than the experimental group at entry. In the two years after the diet phase, there were seven further deaths from carcinoma in the experimental group and ten in the control group. The serum cholesterol level at baseline of men who subsequently died of a carcinoma was not significantly different from that of other men. The authors argued that there was 'no apparent non-dietary explanation' for the higher frequency of cancer in the experimental group. Even though many of the cancer deaths in the experimental group were among men who had not adhered closely to the diet, thus reducing the possibility that the experimental diet caused the excess cancer mortality, they argued that it would be premature 'to make a blanket prescription of a diet high in polyunsaturated fat for the entire population' (Pearce & Dayton, 1971).

Ederer et al. (1971) compared the cancer results from four trials of cholesterol-lowering diets to those of the Veterans Administration (VA) study, which was the first to suggest an excess of cancer mortality. One of the studies included in this review used a cross-over design (Miettinen et al., 1972) which makes it unsuitable for the assessment of cancer risk. None of the trials was designed to examine cancer mortality, and they are thus relatively small: 156 incident cancers and 101 cancer deaths were recorded in all five studies combined, and more than two thirds of these occurred in the elderly population of the VA study. Pooled results from the four other studies suggested relative risks for cancer of 0.75 and 0.62 for morbidity and mortality, respectively, for the experimental diets relative to the control diets, or 1.15 and 1.08 if the VA study was included. None of these risks was significantly different from unity.

The early trials do not therefore provide strong evidence that cholesterol-lowering diets affect cancer risk, either beneficially or adversely, but their limitations need to be remembered. The trials were generally small, and were not sufficiently powerful for detecting an increased cancer risk; they were essentially limited to male subjects, three of them in institutionalized men, and follow-up periods were all less than ten years.

In the Sydney Diet–Heart Study of secondary CHD prevention (Woodhill et al., 1978), 458 men aged 30–59 were studied; 86% had survived one or more heart attacks. They were randomized to receive individual advice either to adopt a low-fat, low-cholesterol diet (221 men), or simply to restrict calories if overweight (237 men), then followed up for up to seven years. The polyunsaturated-to-saturated fat ratio (P:S ratio) was 1.5 in the intervention group and 0.7 in the control group. There was no diet-related difference in CHD mortality. One cancer death occurred in each group.

In the Oslo randomized trial of primary prevention of CHD (Hjermann et al., 1981), 1232 healthy, normotensive men aged 40–49 were selected from among volunteers for a screening programme on the basis of a high serum cholesterol and a high CHD risk score. The intervention consisted of advice to reduce the intake of saturated fat and energy, to increase the intake of polyunsaturated fat and to stop smoking. During the five-year follow-up, the P:S ratio and the reductions in total fat intake, serum cholesterol and tobacco consumption were all substantially greater in the intervention than in the control group. The incidence of fatal and non-fatal myocardial infarction was significantly reduced by 47% in the intervention group, but neither the difference in total mortality (16 versus 24 deaths) nor that in cancer mortality (5 versus 8 deaths) was statistically significant.

One of the largest studies of primary CHD prevention is the Gothenburg trial (Wilhelmsen et al., 1986). In this trial over 30 000 men aged 47–55 years – virtually the entire male cohort in the city born between 1915 and 1925 – were randomized into one intervention group (10 004 men) and two control groups of similar size. The entire intervention group and 2% of one control group were invited to an initial screening examination, with a 75% response. The intervention comprised antihypertensive treatment (if blood pressure exceeded 175/115 mm Hg), clinic-based dietary advice (for the 33.2% whose serum cholesterol exceeded 260 mg per 100 ml) and intensive advice to stop smoking (for the 15.6% smoking more than 15 cigarettes daily). Men with serum cholesterol or smoking habits below these cut-off points were simply given leaflets about diet or smoking. The intervention group and random samples of one control group were re-examined four and ten years later, and complete registration of all deaths was achieved. Risk factor levels decreased markedly over the ten-year follow-up, not only in the intervention group but also among controls; changes in cholesterol were in fact strikingly similar.

All-cause and CHD mortality did not differ significantly between the intervention and control groups. Cancer mortality after a mean follow-up of 11.8 years was slightly lower in the intervention group (3.15%, 315 deaths) than among

Table 2. Randomized trials in which dietary fat modification was a principal intervention: cancer results

Study	Duration[a] (years)	Age range (years)	No. randomized I	No. randomized C	No. of cancers I	No. of cancers C	No. (%) of deaths I		No. (%) of deaths C		Reference
MRC	2-7	<60	199	194			1	(0.5)	6	(3.1)	Research Committee, 1968
Los Angeles VA Trial[a]	10.3	54-88	424	422	64	47	38	(9.0)	27	(6.4)	Pearce & Dayton, 1971
Oslo Diet-Heart	11	30-64	206	206	7	5	1	(0.5)	3	(1.5)	Leren, 1970
Sydney Diet-Heart	2-7	30-59	221	237			1	(0.5)	1	(0.4)	Woodhill et al., 1978
Oslo	5	40-49	604	628			5	(0.8)	8	(1.3)	Hjermann et al., 1981
Gothenburg	10	47-55	10 004	20 018			315	(3.15)	728	(3.64)	Wilhelmsen et al., 1986
MRFIT	7	35-57	6 428	6 438			81	(1.3)	69	(1.1)	MRFIT Research Group, 1982

All the studies in the table involve males only. Data are presented for all cancers (incident and fatal) where available, and include events occurring after the diet phase of the study if available. The Finnish Mental Hospital Study is excluded (see text).

I, intervention group; C, control group

[a] Carcinomas only – 4 other malignancies (sarcoma, leukaemia, etc.) were also recorded in each group; these were discussed separately by the authors and are excluded from this table. The table includes seven and ten post-diet-phase deaths in intervention and control groups, respectively.

the controls (3.64%, 728 deaths). Separate figures were not given for each type of cancer; these might have been of some interest, in view of the numbers of deaths involved.

This randomized study was large, well organized and – unlike almost every other CHD prevention study – population-based, and it would be tempting to treat its negative results as convincing evidence that primary prevention of cancer in the free-living general population is not feasible, at least not with individually-based health education to reduce cholesterol and stop smoking. As the authors point out, however, the period 1970–83 in Sweden saw a general decline in smoking and serum cholesterol in the whole population, and the intervention 'only caused marginal additional effects' in these risk factors. The intervention was mainly directed towards men at the top end of each risk factor distribution, and this would not be expected to produce much change in the mean risk factor levels in the whole group. In the event, mean serum cholesterol fell by about 6% over ten years in both groups, and almost half the smokers stopped smoking, while the rest smoked less, again in both groups. Part of the explanation for this may be that the trial was not only randomized, but involved the entire male population of the city in a given age range. Even though men in the control groups were not aware that they formed part of the study population, ordinary social contacts must have made them aware of the advice being given to the intervention group, and it seems likely that there was considerable contamination of control group behaviour as a result.

There are major differences in design between the Gothenburg study and the MRFIT study (Friedewald *et al.*, this volume), even though both were randomized, controlled trials of primary prevention. In the MRFIT study, 361 662 men who volunteered for screening at 22 different centres in the USA were examined, and the 12 866 men actually randomized were selected from among these volunteers on the basis of their CHD risk factors. Both intervention and control groups were then individually followed up. The MRFIT study thus provides information about the effects of intensive individual health education in a motivated subgroup of the population – men who volunteered for CHD screening and then discovered that they were at high risk of the disease. In Gothenburg, by contrast, the entire male population aged 47–55 was first randomized: 25% of the intervention group did not participate in the initial screening, or in the intervention, and the majority of the 20 000 controls were unaware of either their inclusion in the study or the level of their CHD risk factors. Results are nevertheless presented on an 'intention-to-treat' basis, as specified by the trial design. This study therefore provides information about the effects of an intervention in the general population, diluted by the non-participation of a substantial minority of men with less favourable health, and whose mortality was higher than that of participants for each cause of death.

If the Gothenburg trial had achieved a reduction in both risk factors and mortality, its results would have been particularly persuasive, because of its population-based design and the inclusion of non-participants in the analysis. The implication for trials designed to test the efficacy of an intervention for primary prevention of diseases like CHD and cancer, however, seems to be that a selected population is likely to be more rewarding.

While the Gothenburg trial failed to achieve a clear difference in risk factors between the intervention and control groups, and showed no effect of the intervention on either CHD or overall cancer mortality, it did provide clear evidence that 'whole populations can and do change their [diet and smoking] habits substantially' over a decade (Rose, 1986). This is relevant because effective primary prevention must depend not only on the efficacy of the intervention but on the likelihood of its being adopted by the general population.

Conclusions

Dayton et al. (1969), commenting on one of the early trials before the weight of evidence on CHD prevention was clear, summed up the difficulty which still confronts us now with regard to cancer: '... additional studies of the question will add little to the dozen or so already reported unless they are bigger and better and longer. The chief desiderata are experimental designs which will relieve us of concern about bias and about confounding effects, and study populations large enough to yield results at high confidence levels and with reasonable hope of demonstrating significant differences in total mortality, if such differences exist.'

The reason for the lack of evidence from cancer primary prevention trials is essentially practical: such trials would be extremely demanding to carry out. Zelen (1988) has even argued that such trials are not feasible, at least if the main endpoint to be studied is cancer incidence, but that trials for which some intermediate endpoint (such as a biological marker of disease) could be found would still be feasible. Byar (1988) illustrates the difficulties by comparing trials of cancer prevention and treatment. Since most subjects in a prevention trial will not develop cancer, the number of subjects required for such a trial may be 100 or more times greater than for a conventional treatment trial. Similarly, since the interval between treatment and relapse or death (the usual endpoints in treatment trials) is often fairly short, such trials can be completed quickly, whereas the latent period between a dietary change and the occurrence (or prevention) of a cancer is likely to be five or ten years or more. Prevention trials must therefore last for at least that long to have a chance of success, which implies the need for subjects to adhere to the dietary change over a long period, and for their compliance to be monitored. Zelen (1988) has pointed out that the loss of statistical power associated even with relatively good compliance, e.g. 80%, may be disproportionately severe (depending on the square of the compliance). The difficulty of following up a large population of healthy subjects is also much greater than that of following up a smaller group of patients who have a direct interest in returning to see their doctor. All these requirements imply considerable expense, and thus the need both for a careful justification of the dietary hypothesis to be tested and for a pilot study to demonstrate successful recruitment of subjects and their compliance with the intervention. Despite the potential benefit of discovering a feasible dietary approach to the primary prevention of a common cancer such as breast or colon cancer, these difficulties would appear so far to have prevented the execution of any satisfactory study.

References

Broitman, S.A., Vitale, J.J., Vavrousek-Jakuba, E. & Gottlieb, L.S. (1977) Polyunsaturated fat, cholesterol and large bowel tumorigenesis. *Cancer*, *40*, 2455-2463

Byar, D.P. (1988) The design of cancer prevention trials. In: Scheurlen, H., Kay, R. & Baum, M., eds, *Cancer Clinical Trials: A Critical Appraisal* (Recent Results in Cancer Research, Vol. 111), Berlin, Heidelberg, New York, Springer-Verlag, pp. 34-48

Clifford, C.K., Butrum, R.R., Greenwald, P. & Yates, J.W. (1986) Clinical trials of low fat diets and breast cancer prevention. In: Ip, C., Birt, D.F., Rogers, A.E. & Mettlin, C., eds, *Dietary Fat and Cancer*, New York, A.R. Liss, pp. 93-115

Dayton, S., Pearce, M.L., Goldman, H., Harnish, A., Plotkin, D., Shickman, M., Winfield, M., Zager, A. & Dixon, W. (1968) Controlled trial of a diet high in unsaturated fat for prevention of atherosclerotic complications. *Lancet*, *ii*, 1060-1062

Dayton, S., Pearce, M.L., Hashimoto, S., Dixon, W.J. & Tomiyasu, U. (1969) A controlled clinical trial of a diet high in unsaturated fat in preventing complications of atherosclerosis. *Circulation*, *40* (Suppl. II), 1-63

Ederer, F., Leren, P., Turpeinen, O. & Frantz, I.D. (1971) Cancer among men on cholesterol-lowering diets: experience from five clinical trials. *Lancet*, *ii*, 203-206

Gammal, E.B., Carroll, K.K. & Plunkett, E.R. (1967) Effects of dietary fat on mammary carcinogenesis by 7,12-dimethylbenz(α)anthracene in rats. *Cancer Res.*, *27*, 1737-1742

Greenwald, P., Cullen, J.W. & McKenna, J.W. (1987) Cancer prevention and control: from research through applications. *J. Natl Cancer Inst.*, *79*, 389-400

Hjermann, I., Velve Byre, K., Holme, I. & Leren, P. (1981) Effect of diet and smoking intervention on the incidence of coronary heart disease: report from the Oslo Study Group of a randomised trial in healthy men. *Lancet*, *ii*, 1303-1310

Leren, P. (1966) The effect of plasma cholesterol lowering diet in male survivors of myocardial infarction: a controlled clinical trial. *Acta Med. Scand.*, Suppl. 466

Leren, P. (1970) The Oslo Diet-Heart study: eleven-year report. *Circulation*, *42*, 935-942

Miettinen, M., Turpeinen, O., Karvonen, M.J., Elosuo, R. & Paavilainen, E. (1972) Effect of cholesterol-lowering diet on mortality from coronary heart-disease and other causes: a twelve-year clinical trial in men and women. *Lancet*, *ii*, 835-838

MRFIT Research Group (1982) Multiple risk factor intervention trial: risk factor changes and mortality results. *J. Am. Med. Assoc.*, *248*, 1465-1477

Pearce, M.L. & Dayton, S. (1971) Incidence of cancer in men on a diet high in polyunsaturated fat. *Lancet*, *i*, 464-467

Reddy, B.S., Narisawa, T., Vukusich, D., Weisburger, J.H. & Wynder, E.L. (1976) Effect of quality and quantity of dietary fat and dimethylhydrazine in colon carcinogenesis in rats. *Proc. Soc. Exp. Biol. Med.*, *151*, 237-239

Research Committee (1968) Controlled trial of soya-bean oil in myocardial infarction: report of a research committee to the Medical Research Council. *Lancet*, *ii*, 693-700

Rose, G. (1986) Comment on the Göteborg study by Wilhelmsen et al. *Eur. Heart J.*, *7*, 288

Rose, G., Hamilton, P.J.S., Colwell, L. & Shipley, M.J. (1982) A randomised controlled trial of anti-smoking advice: 10-year results. *J. Epidemiol. Commun. Health*, *36*, 102-108

Thompson, D. (1987) The Women's Health Trial: a randomized trial to prevent breast cancer incidence with a low fat diet. Presentation no. 3: results of the feasibility phase (Abstract 507). *International Epidemiological Association, 11th Scientific Meeting*, Helsinki, Finland, 8-13 August 1987

Wilhelmsen, L., Berglund, G., Elmfeldt, D., Tibblin, G., Wedel, H., Pennert, K., Vedin, A., Wilhelmsson, C. & Werkö, L. (1986) The multifactor primary prevention trial in Göteborg, Sweden. *Eur. Heart J.*, *7*, 279-288

Woodhill, J.M., Palmer, A.J., Leelarthaepin, B., McGilchrist, C. & Blacket, R.B. (1978) Low fat, low cholesterol diet in secondary prevention of coronary heart disease. *Adv. Exp. Med. Biol.*, *109*, 317-330

Zelen, M. (1988) Are primary cancer prevention trials feasible? *J. Natl Cancer Inst.*, *80*, 12-15

WHO EUROPEAN COLLABORATIVE TRIAL IN THE MULTIFACTORIAL PREVENTION OF CORONARY HEART DISEASE

World Health Organization European

Collaborative Group[1]

The WHO European Collaborative Trial in the multifactorial prevention of coronary heart disease (CHD) has been described in full (WHO European Collaborative Group, 1974), and its interim results (WHO European Collaborative Group, 1980, 1982, 1983) and final results (WHO European Collaborative Group, 1986; Rose, 1987) were published. The present paper corresponds to the specific interests of the workshop. This explains, on one hand, the brevity of certain descriptions and, on the other hand, the complete omission of such important issues as standardization and data processing.

The objectives of the study were to find answers to the questions of how much the main coronary risk factors could be reduced in populations and how such changes affect the incidence of fatal CHD, total CHD (fatal CHD combined with non-fatal myocardial infarction), and total mortality. Since this was to be a trial of a realistic public health policy, resources for intervention were limited to a level that could be implemented widely: overall, intervention staff averaged two doctors and one nurse/nutritionist per 8000 intervention subjects. Thus, the trial was not primarily a test of the scientific hypothesis that risk is reversible, but of the effects of a particular preventive policy.

The design of the study is outlined in Figure 1. One of each matched pair of factories was randomized to intervention and the other served as control, the members of each pair being sufficiently separated to avoid contamination. Risk factor and disease status at entry was characterized by screening all the intervention men (which also identified those at high risk) but only a 10% sample of controls. Incidence measurement in the control population was confined to the 90% who had no contact with the study, since it was hypothesized that the behaviour of the other 10% could have been influenced by the screening itself.

---

[1] Members of the Group: Belgium: Dr G. de Backer, Dr M. Kornitzer, Michele Dramaix, Prof. C. Thilly (data-coordinator); Italy: Prof. A. Menotti, Prof. G. Ricci, Prof. G.C. Urbinati, Prof. G. Farchi; Poland: Prof. S.I. Rywik, Prof. J. Sznajd, Prof. W.B. Szostak, Dr M. Magdon, Dr J. Charzewska; Spain: Prof. I. Balaguer-Vintro, Dr. L. Tomas-Abadal, Dr S. Sans; UK: Prof. G. Rose (chairman), Prof. H.D. Tunstall Pedoe, Dr R.F. Heller, Mr M.J. Shipley; World Health Organization (European Office): Dr G. Lamm

Figure 1. Outline of the experimental design

	INTERVENTION	CONTROL
1. Randomize		
2. Screen	All	Random 10%
3. Treat	Cholesterol-weight-cigarettes-activity-BP	----
4. Monitor change	5% annually	10% biennially
5. Incidence	Cardiovascular disease + other deaths	As in intervention group
6. Final examination	All still employed	As in intervention group

The study group comprised all employees aged 40 to 59 in the participating factories. The aim was to evaluate prevention in total communities (occupationally defined).

The intervention programme combined both a population approach (health education for all) and a 'high risk' approach (screening, followed by individual care). Men with the highest coronary risk scores were counselled and followed individually by doctors, nurses and nutritionists. All men received health education material, such as letters and specially prepared booklets, at intervals throughout the project, and many subjects were also seen individually. Special posters were displayed widely, appropriate menus were available in canteens, and (where possible) talks, film shows and discussions were held for men and their wives.

The intervention measures were dietary advice to reduce serum cholesterol, stopping smoking, weight loss (for those 15% or more overweight), daily exercise (for the sedentary) and drug and hygienic treatment of hypertension (if systolic pressure averaged 160 mmHg or above).

Risk-factor changes in the trial were monitored in intervention men by examining each year a fresh 5–10% sample, while in controls the same 10% of men were re-examined every two years. All survivors still in employment were invited to a final examination after 5 or 6 years in the trial, to estimate the final differences in risk-factor levels.

Survival status at the end of the trial was established in over 99% of subjects. The registration of critical events was based on death certificates and a review of the clinical records.

Evaluation design

Each factory pair yielded (for each critical event category) one independent estimate of the effect of intervention. These estimates, each weighted by the inverse of its variance, were combined to yield normally distributed statistics from which we obtained significance levels and confidence intervals.

Risk factor change was obtained from the calculation for each individual of the multiple logistic function (MLF) change from entry to the 2-year, the 4-year, and the final examination. Individual changes were then averaged to estimate the overall net change for each intervention factory.

For the particular model used in the analysis, the effect of intervention in each factory pair was expressed by the logarithm of the ratio of the intervention to the control rate. (This assumes that in each factory pair the percentage change in incidence rate due to intervention is independent of the magnitude of the underlying rates.) The rates actually observed in each factory are only estimates of the true underlying event rates, and since the study has randomized whole factories rather than individuals, the analysis must allow for variation in their underlying event rates. The relation of outcome to risk-factor change was therefore assessed by means of a modification of the maximum likelihood regression method of Pocock et al. (1981):

$$\log_e(P_i/P_c) = a + b\,(\Delta\text{MLF}) + \epsilon,$$

where P_i = rate in intervention factory,

P_c = rate in control factory,

a = a constant (represents any effect of intervention for zero net change in risk factors, as measured),

ΔMLF = net mean change in risk factors (summarized by a multiple logistic function),

b = slope of the regression (estimates the strength of the relation between risk factor change and outcome),

and ϵ = error term.

This analysis, however, does not take account of risk-factor levels in individual factories at entry to the study. Underlying differences within factory pairs could be large, but their effect on event rates can be allowed for. The relation between the net differences in incidence and net changes in MLF, incorporating an adjustment for entry MLF, was examined for each factory pair as follows:

$$P_i - P_c = b_1\,(\text{MLF}_i - \text{MLF}_c) + b_2\,\Delta\text{MLF} + \epsilon,$$

where

i = intervention factory,

c = control factory,

P = rate,

MLF = entry MLF,

ΔMLF = net mean change (intervention–control) in multiple logistic summary during trial,

ϵ = error term,

and b_1, b_2 = slopes of regression

This function was fitted by using a modification of the maximum likelihood regression method of Pocock et al. (1981).

This method was also used to estimate the overall effect of intervention, ignoring risk-factor changes. It provides tests of significance that take account of the randomization of factories rather than individuals. In this case, the regressions simplify and become equivalent to weighted means of the change in incidence.

Main results

The trial enrolled 60 881 men in 80 factories in the four countries for which incidence data are represented, and 2851 men in Spain.

The effects of intervention have been expressed for each examination by the net difference between the mean of individual changes from baseline in intervention men and the corresponding mean in controls. The overall changes (Table 1) are given as an unweighted mean over all examinations (which takes no account of a possible latency of effects). The combined effect of individual risk-factor changes is summarized by a multiple logistic function estimate of predicted CHD incidence, using coefficients based on the experience of the European cohorts of the Seven Countries Study (G. Farchi and A. Menotti (1976), unpublished).

Table 1. Net percentage changes in risk factors, averaged over all examinations

Risk factor	Belgium (1)	Italy (2)	Poland (3)	UK (4)	All (1-4)	Spain
Cholesterol	-0.9	-4.8	-1.0	-0.4	-1.2	-4.3
Smokers %	-1.7	+2.3	+1.3	-4.5	-1.9	+4.8
Cigarettes/day	-3.7	-5.5	-5.9	-15.6	-8.8	+2.3
Weight	+0.2	-1.9	-1.0	-0.4	-0.4	-2.0
Systolic BP	-2.3	-4.1	-0.6	-1.6	-2.0	+4.0
Combined (MLF)	-15.8	-28.2	-19.0	-3.9	-11.1	-0.4

Changes are modest, but nearly all are in the right direction and statistically very significant. The multiple logistic summary predicts a fall of 11% in CHD incidence, if risks were fully and immediately reversible. This figure ranged from 28.2% in the Italian centre (where the ratio of intervention staff to subjects was highest) down to 3.9% in the UK (where the ratio was lowest).

Table 2 shows that at each stage of the trial the more vigorous personal intervention effort among high-risk men produced much bigger risk-factor changes than the general education campaign in the study group as a whole. The changes in the high-risk men were of the same order as those obtained in the US Multiple Risk Factor Intervention Trials (MRFIT) study (1982). Effects were greatest at the four-year point of the trial.

The number of critical events was lower in intervention than in control groups; the differences were 6.9% for fatal CHD, 14.8% in non-fatal myocardial infarction, 10.2% for 'total' CHD, and 5.3% for total deaths (Table 3). Mortality from causes other than CHD was almost identical in the two groups. This result is reassuring with respect to the safety of the intervention programme: it does not suggest that a reduction in deaths from heart attack is obtained at the expense of an increase in risk of death from other causes.

Table 2. Net percentage changes in a multiple logistic estimate of CHD risk at various stages of the trial

	Net % change in a multiple logistic estimate			
	2 year	4 year	Final	Total
High-risk men	−18.2	−24.7	−15.8	−19.4
All men	−12.5	−15.1	−7.0	−11.1

Table 3. Critical events: numbers, net percentage difference between intervention and control groups, and 95% confidence interval

	No.	% Difference	95% Confidence interval
Fatal CHD	826	−6.9 ($p = 0.8$)	−19 to +7
Non-fatal MI	1010	−14.8 ($p = 0.06$)	−28 to +1
Total CHD[a]	1800	−10.2 ($p = 0.07$)	−20 to +1
All deaths	2511	−5.3 ($p = 0.4$)	−15 to +6

[a] Men with fatal and non-fatal events count once only CHD, coronary heart disease; MI, myocardial infarction

Within any one factory pair the reduction in fatal CHD was significantly related to its reduction in risk factors (measured by the MLF): advice, therefore, is effective to the extent that it is accepted.

Table 4 shows the estimates of slope obtained, together with their standard errors, for various end-points. The estimates in the first column indicate that, except for non-fatal myocardial infarction, imbalance in risk factors at entry between intervention and control factories was significantly associated with their six-year event rate differences. The table also shows estimates of the effect of intervention on the six-year event rate in relation to compliance, after taking into account any entry imbalance of risk factors. The results are significant at the 5% level for each of the trial's predefined end-points: fatal myocardial infarction, total CHD, and total mortality. This confirms that our advice on CHD reduction was effective, to the extent that it was accepted. The benefit includes a significant reduction in total mortality.

Before presenting the findings on cancer mortality, it is stressed that the figures should be interpreted with caution. The present trial was planned and executed as an intervention trial on coronary heart disease. Cancer mortality was registered as part of the mandatory analysis in any such trial of all-causes mortality. In other words, efforts were made to ascertain reliably non-CHD mortality, but within this category no special efforts were made to investigate its subgroups. If the cause of death was available, this was entered on the patient's file; if not, the only investigation involved finding out whether or not the cause of death was due to CHD. This policy was strictly followed in Belgium, so no information is given here on their cancer mortality. Due to this secondary interest in cancer deaths, the rates in the other centres should be observed with reservation.

Table 4. Regressions of change in incidence[a] (P_1-P_c) on initial difference in risk (initial MLF_1–initial MLF_c) and differences in compliance (MLF risk factor change)

Outcome	Differences in MLF at entry		Differences in compliance	
	Slope	Standard error (SE)	Slope	Standard error (SE)
Fatal CHD	0.65[b]	(0.25)	0.60[b]	(0.25)
Non-fatal MI	0.72	(0.40)	0.64	(0.45)
Total CHD	1.23[b]	(0.46)	1.18[b]	(0.53)
Total deaths	1.98[b]	(0.56)	1.41[b]	(0.60)
Non-CHD deaths	1.18[b]	(0.49)	0.63	(0.52)

[a] Incidence rate and MLF score are both per person per 6 years
[b] $p < 0.05$ (two-tailed test)
CHD, coronary heart disease; MI, myocardial infarction

Tables 5–8 present the cancer mortality rates per 1000 in six years in the UK, Poland, Italy and Spain. Age standardization was done by the direct method using the whole trial population as basis. Rates are given separately for subjects in the treatment and control factories, with significance levels calculated by the two-tailed test for any differences found.

Table 5. Cancer mortality rates (age-standardized) in the United Kingdom (numbers of deaths in parentheses)

Primary site	Treatment No. = 9724		Control No. = 8476		p-level of difference
Digestive:					
Stomach	0.3	(31)	0.3	(25)	
Colon, rectum, pancreas	0.1	(13)	0.1	(10)	
Respiratory:					
Lung	1.2	(137)	1.0	(93)	$p = 0.16$
Other	1.2	(119)	1.0	(95)	
All cancer	2.7	(300)	2.5	(223)	$p = 0.24$

Table 6. Cancer mortality rates (age-standardized) in Poland (numbers of deaths in parentheses)

Primary site	Treatment No. = 9110		Control No. = 8118		p-level of difference
Digestive:					
Stomach	0.3	(25)	0.3	(22)	
Colon, rectum, pancreas	0.1	(10)	0.1	(10)	
Respiratory:					
Lung	0.5	(42)	0.5	(33)	$p = 0.70$
Other	0.9	(69)	0.6	(44)	
All cancer	1.8	(146)	1.5	(109)	$p = 0.20$

Note: Some denominators differ from those used for crude rates, because age is missing for some subjects

Table 7. Cancer mortality rates (age-standardized) in Italy (numbers of deaths in parentheses)

Primary site	Treatment No. = 3131		Control No. = 2896		p-level of difference
Digestive:					
Stomach	0.2	(8)	0.4	(11)	
Colon, rectum, pancreas	0.3	(9)	0.4	(10)	
Respiratory:					
Lung	1.1	(36)	0.8	(23)	$p = 0.26$
Other	1.4	(46)	1.5	(44)	
All cancer	3.0	(99)	3.1	(88)	$p = 0.68$

Note: Some denominators differ from those used for crude rates, because age is missing for some subjects

Emphasizing once more that influence on and analysis of cancer mortality was not the declared intention of the trial, the overall experience in the treatment and control factories in the four countries are presented in Table 9.

If the results are evaluated given the cautions discussed above, two conclusions may be drawn: firstly, anti-smoking intervention appears to be insufficient to influence lung cancer mortality rates in the intervention factories, even in the UK, where anti-smoking advice appeared to be the most successful. Whether this is due to small numbers, insufficient level of intervention or a too short observation period, cannot be answered from our data. Secondly, the cheap and feasible intervention applied achieved a significant reduction in the incidence of CHD and total mortality without any apparent increase of cancer mortality. In view of the still lingering problem of low cholesterol levels and cancer, this appears to be an important public health message.

Table 8. Cancer mortality rates (age-standardized) in Spain (numbers of deaths in parentheses)

Primary site	Treatment No. = 1383		Control No. = 1467		p-level of difference
Digestive:					
Stomach	0.1	(2)	0.4	(7)	
Colon, rectum, pancreas	0.1	(2)	0.3	(4)	
Respiratory:					
Lung	0.3	(4)	0.5	(7)	$p = 0.40$
Other	0.7	(11)	1.3	(18)	
All cancer	1.3	(19)	2.4	(36)	$p = 0.02$[a]

[a] Significant

Table 9. Cancer mortality rates (age standardized) for entire World Health Organization European trial

Primary site	Treatment No. = 23 358		Control No. = 20 957		Z-Test of difference
Digestive:					
Stomach	0.3	(66)	0.3	(65)	-0.727
Colon, rectum, pancreas	0.1	(34)	0.2	(34)	-0.565
Respiratory:					
Lung	0.9	(219)	0.8	(156)	1.848 ($p = 0.064$)
Other	1.0	(245)	1.0	(201)	0.670
All cancer	2.4	(564)	2.2	(456)	1.167

Note: None of the differences reported reaches the conventional level of significance

The limitations in the use of obtained information on cancer in a purely cardiovascular study are merely one reason to call for intensified collaboration between various medical specialities in the future. The assumed causative role of many environmental and lifestyle characteristics for a number of chronic diseases, the largely uniform epidemiological framework of these studies and the expected ability for improved insight into the problem of competing risk are additional strong arguments for a reasonable multidisciplinary approach.

References

Multiple Risk Factor Intervention Trial Research Group (1982). Multiple risk factor intervention trial. Risk factor changes and mortality results. *J. Am. Med. Assoc.*, 248, 1465-1477

Pocock, S.J., Cook, D.G. & Beresford, S.A.A. (1981) Regression of area mortality rates on explanatory variables: what weighting is appropriate? *Appl. Statistics*, 30, 286-295

Rose, G. (1987). European collaborative trial of multifactorial prevention of coronary heart disease [letter]. *Lancet*, i, 685

World Health Organization European Collaborative Group (1974). An international controlled trial in the multifactorial prevention of coronary heart disease. *Int. J. Epidemiol.*, 3, 219

World Health Organization European Collaborative Group (1980). Multifactorial trial in the prevention of coronary heart disease: 1. Recruitment and initial findings. *Eur. Heart J.*, 1, 73-80

World Health Organization European Collaborative Group (1982). Multifactorial trial in the prevention of coronary heart disease: 2. Risk factor changes at two and four years. *Eur. Heart J.*, *3*, 184-190

World Health Organization European Collaborative Group (1983). Multifactorial trial in the prevention of coronary heart disease: 3. Incidence and mortality results. *Eur. Heart J.*, *4*, 141-147

World Health Organization European Collaborative Group (1986). European collaborative trial of multifactorial prevention of coronary heart disease: final report on the 6-year results. *Lancet*, *i*, 869-872

Evaluating Effectiveness of Primary Prevention of Cancer
Ed. M. Hakama, V. Beral, J.W. Cullen & D.M. Parkin
Lyon, International Agency for Research on Cancer
© IARC, 1990

CHANGES IN CANCER INCIDENCE IN NORTH KARELIA, AN AREA WITH A COMPREHENSIVE PREVENTIVE CARDIOVASCULAR PROGRAMME

T. Hakulinen[1,2], E. Pukkala[1], M. Kenward[3], L. Teppo[1],
P. Puska[4], J. Tuomilehto[4] and K. Kuulasmaa[4]

[1]*Finnish Cancer Registry, Helsinki, Finland*
[2]*Department of Statistics, University of Helsinki, Helsinki, Finland*
[3]*Department of Applied Statistics, University of Reading, Reading, UK*
[4]*Department of Epidemiology, National Public Health Institute, Helsinki, Finland*

Introduction

In the 1960s, the world's highest cardiovascular disease (CVD) mortality rates were recorded among men in Finland (Ovcarov & Bystrova, 1978). Within Finland, the rates were highest in the eastern parts of the country, in particular, in the easternmost county of North Karelia. This fact prompted the North Karelia Project, a comprehensive community-based programme to prevent CVD in that county (Puska *et al.*, 1981). The project was formulated in close cooperation with Finnish experts and the WHO, and was launched in 1972.

The main objective of the North Karelia Project was to decrease the mortality and morbidity rates of major cardiovascular diseases through a general reduction in smoking, serum cholesterol concentration and high blood pressure. To achieve this, a comprehensive community-based intervention programme aiming at changes in smoking habits and diet and based on local community action and local service structures was implemented in North Karelia in 1972 (Puska *et al.*, 1981).

Large independent cross-sectional population samples were surveyed in 1972, 1977 and 1982 in North Karelia and in a reference area, the county of Kuopio (adjacent to North Karelia), to assess changes in risk factors and health behaviour. Smoking in males, serum cholesterol concentration and blood pressure levels declined among the middle-aged population in North Karelia more than in the reference area (Puska *et al.*, 1983). The changes in the serum cholesterol level were probably a consequence of dietary changes, especially of a reduction in the intake of saturated fat.

Many of the risk factors in cardiovascular disease have also been considered as risk factors for cancer. The link between smoking and cancer at several sites – lung, larynx, oesophagus, oral cavity, pharynx, bladder, lip, and pancreas – is well established (IARC, 1986). The association between fat, fibre and salt intake and several cancer types has been a subject of intensive research (Armstrong & Mann, 1985). Most hypotheses concerning dietary risk factors in CVD and cancer suggest associations in the same direction, but the possibility of opposite effects has also been discussed. Cancers that have been related to diet in epidemiological studies are those of the gastrointestinal tract, breast and corpus uteri (Armstrong & Mann, 1985; Hill, 1985).

The existence of a comprehensive national cancer registration system in Finland since 1953 (Saxén & Teppo, 1978) provided an opportunity to examine changes in the cancer incidence in North Karelia, in the reference area of Kuopio county, and in the rest of Finland. The purpose of this study was to ascertain whether the decreased levels of CVD risk factors observed in North Karelia had also had an effect on the cancer incidence observed in that county.

Material and methods

Population surveys

In 1982, North Karelia comprised 3.7% (177 000) of the population of Finland. The corresponding proportion for Kuopio county was 5.2% (253 000). Three population surveys were carried out to assess changes in certain CVD risk factors in North Karelia and Kuopio counties, one at the outset of the project in 1972, and two at five-year intervals thereafter. Independent random samples were drawn from the national population register and surveyed using strictly standardized methods each time. The surveys included a questionnaire on demographic data, socioeconomic status, medical history and health behaviour, and measurements of height, weight and blood pressure (Puska et al., 1983). A venous blood specimen was taken for determination of the serum cholesterol concentration.

The participation rates were 94%, 89% and 80% in North Karelia in 1972, 1977 and 1982 respectively. The corresponding rates for the county of Kuopio were 91%, 91% and 82%. The present analysis is limited to subjects aged 30–59 years at the time of survey. Table 1 shows the number of males and females studied in the three surveys in the counties of North Karelia and Kuopio.

Table 1. Number of males and females aged 30–59 years studied in the three population surveys in North Karelia and in the county of Kuopio

Year of survey	North Karelia		County of Kuopio	
	Males	Females	Males	Females
1972	1834	1973	2665	2769
1977	1785	1845	2616	2756
1982	1250	1285	1197	991

Cancer incidence

The Finnish Cancer Registry has data on all cancer cases diagnosed in Finland since 1953. It is population-based and covers the whole of Finland. On the basis of a code for the place of residence (municipality) of the patient at the time of diagnosis of cancer, and population data by municipality drawn from the files of the Central Statistical Office, incidence rates were calculated for North Karelia, Kuopio county and for all the other counties in Finland combined.

Only cancers that are fairly common and possibly related to smoking and/or diet were chosen for analysis, i.e., cancers of the lung, larynx, bladder, stomach, colon, rectum, pancreas, oesophagus, lip, breast and corpus uteri. The analysis was limited to adults in the age range 30–74 years. For statistical modelling, the lower age limit had to be raised to 35 years (colon, pancreas, corpus uteri), 40 years (bladder, rectum), 45 years (oesophagus, lip) or 50 years (larynx) owing to insufficient numbers of cases in the lower age groups. For the same reason, breast cancer in males and laryngeal cancer in females were totally excluded from the analyses. The cancer incidence data used for the analysis covered the 30-year period 1957–86.

Statistical methods

For economy of analysis and to make it possible by age, period and synthetic cohort (Clayton & Schifflers, 1987), the cancer incidence data were grouped into six five-year calendar periods and five-year age groups. Log-linear age–period–cohort–region models (Clayton & Schifflers, 1987) were fitted for each of the 19 sex–site combinations separately. Cumulative incidence rates (Waterhouse et al., 1982) were used to describe incidence rates over broader age ranges.

The idea was to compare the observed incidence rates in North Karelia in the post-programme years, 1977–86, with the expected values that would have been derived according to past experience in the years 1957–76 in North Karelia itself, and in the whole 30-year period, 1957–86, in the two other areas, Kuopio county and the rest of Finland. Basically the incidence rates in North Karelia for 1977–86 were treated as if they were missing and could be predicted from the rest of the data available (i.e., prediction basis). Allowing a time lag for the effects to emerge, the most recent five-year period, 1982–86, was chosen as the main target of the evaluation.

The models used in this exercise should fit fairly well with the data that form the prediction basis. The estimates and fit of the models were investigated using the GLIM statistical programme package (Payne, 1986). With the models, predicted incidence rates and numbers of cases were calculated for all observational units, including North Karelia, in 1957–86.

A base model for prediction included the linear trend ('drift', see Clayton & Schifflers, 1987), the square of the numerical code (from one to six) for the period, the square and cube of the numerical code (consecutive integers starting from one) for cohort, age as a categorical variable (five-year groups) and county as a three-level categorical variable (North Karelia, Kuopio, others). Allowance was made for the different incidence trends in the three areas by including an interaction between drift and area. A more complicated model was needed only for four sex–site combinations. For bladder cancer in both sexes, a categorization of the period was used instead of the square term. For cancer of the corpus uteri, categorical cohort and period variables were used instead of the corresponding power terms. The fits of the models were markedly improved in this way but the effect on the predicted values was rather small. For confirmatory purposes, all the other analyses were repeated with a categorical period variable instead of a second power numerical variable. The fits did not markedly improve, and the predicted values were practically unaffected. No significant interactions between area and age or area and cohort were observed.

Lung cancer in males was a special problem. A model with a categorical period and cohort gave a deviance (119.1) close to the degrees of freedom (114), and all the simpler models were rejected. A visual comparison of the fitted and observed values, however, indicated systematic deviations for North Karelia (Figure 1) due to the major weight on the other, more populous counties in the fit. The figure remained practically unchanged even when a full age–period interaction (including thus cohort effects) was included in the model. Separate models were then fitted for each of the three areas. The one giving a fairly good fit for North Karelia in 1957–76 included a drift, no period term, a second-power numerical cohort term and age as a categorical variable. This model was then extrapolated for the years 1977–86 to give the predicted values.

Indicator variable techniques for periods were used in the age–period–cohort–region models to assess the significance of interactions between area and five-year periods in 1977–86. For this analysis, the North Karelian data from the last ten-year period were also included. This test gives information on differences in the temporal development of incidence rates between areas during the ten-year period of interest. Moreover, indicator variables for the period were used to investigate whether the last ten years had introduced a significant change in the incidence trend observed in North Karelia.

Results

Changes in risk factors

During the ten-year survey period the prevalence of smokers in males decreased in North Karelia from 52% to 39% (Table 2). A smaller decrease, from 50% to 45%, was observed in Kuopio county. The difference in the prevalence of smoking between the counties in 1982 was confirmed by analyses of the serum thiocyanate levels. The prevalence of smoking among women increased both in North Karelia and in Kuopio county, mainly because new cohorts with a higher prevalence of smokers reached the age groups in question.

Several indicators that describe fat intake showed a decline in North Karelia. Similar but, in general, smaller changes were observed in the reference area. Fat intake from milk and butter, representing about half the total fat intake in the Finnish population, decreased in North Karelia by 35% in males and 36% in females during the period 1972–82 (Table 2). In the county of Kuopio the corresponding decreases were 15% and 23%, respectively. The mean cholesterol level decreased in both areas, among men somewhat more in North Karelia. For further details of the changes, see Puska *et al.* (1983).

Figure 1. The observed and predicted lung cancer incidence rates by birth-year cohort and age in North Karelian males in 1957–86.

The predicted rates are based on one log-linear model for the incidence in three areas with categorical age, period, cohort and area variables, drift and interaction between area and drift (for details, see methods section).

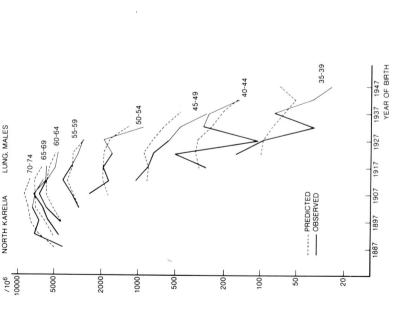

Figure 2. The observed and predicted lung cancer incidence rates by birth-year cohort and age in North Karelian males in 1957–86.

The predicted rates are based on a log-linear model for the incidence in North Karelia with categorical age, numerical cohort code squared and drift (for details, see methods section).

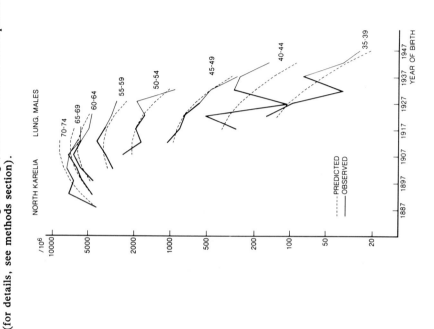

Table 2. Changes in main health behaviour and risk factors in North Karelia (NK) and in the county of Kuopio (KU) according to cross-sectional population surveys in 1972, 1977 and 1982 among men and women aged 30-59 years

Sex and year	Percentage of smokers		Amount of smoking[a]		Fat from milk and on bread/day (g)		Serum cholesterol (mmol/l)	
	NK	KU	NK	KU	NK	KU	NK	KU
Males								
1972	52	50	10.0	8.6	83	72	7.1	6.9
1977	44	45	8.6	8.5	55	62	6.7	6.8
1982	39	45	6.6	7.9	54	61	6.3	6.3
Females								
1972	10	11	1.1	1.2	45	39	7.0	6.8
1977	10	12	1.2	1.4	30	33	6.6	6.5
1982	18	18	1.7	1.9	29	30	6.2	6.0

[a] Number of cigarettes, cigars and pipes per day per subject (smokers and non-smokers taken together)

Changes in cancer incidence

Lung. The observed incidence of lung cancer in North Karelian males in 1977-86 was less than that predicted in ages 65-74 years (Figure 2). In other age groups the observed rates were closer to those predicted. In 1977-86, there was significant ($p < 0.05$) heterogeneity in the pattern of the incidence curves between the three areas, and in ages 55-74 the rates both in Kuopio county and in North Karelia seemed to come somewhat closer to the rates in the other counties (Figure 3). In younger males, aged 30-54, there was comparatively little heterogeneity between the rates in the last five-year period, 1982-86. The observed number of lung cancer cases in 1982-86 in North Karelian males was 362, which is about 10% smaller than that predicted, 406.8 (Table 3). In females, the observed numbers of lung cancer slightly exceeded those predicted (Table 4).

Larynx. There was significant ($p < 0.01$) heterogeneity in the pattern of the incidence curves of laryngeal cancer in males between the areas in 1977-86. Both in North Karelia and in Kuopio county the rate in the second half of the period was much higher than in the first half (Figure 4). However, the observed and predicted incidence rates of laryngeal cancer in North Karelian males aged 55-75 were very close to each other in 1982-86.

Stomach. The observed incidence of stomach cancer in North Karelian males in 1977-86 exceeded that predicted in ages 55-74 years (Figure 5). The pattern of the observed incidence curve in 1977-86 was significantly ($p < 0.05$) different from that in the previous years. In 1982-86 there was an excess in the number of observed cancer cases in ages 55-74 compared with that predicted in both sexes (Tables 3 and 4).

Figure 3. The observed and predicted lung cancer cumulative incidence rates in males aged 30–54 and 55–74 years in North Karelia (NK), Kuopio county (KU) and other counties in Finland (OC) in 1957–86 by calendar period and the predicted rates (x) for North Karelia

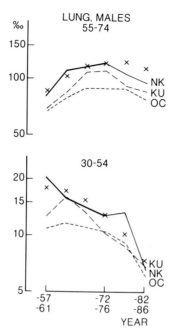

Table 3. Observed (Obs.) and predicted (Exp.) numbers of cancer cases in North Karelian males in 1982–86 at selected sites by age group

Site	Lowest age (LO)	Ages				Total	
		LO–54		55–74			
		Obs.	Exp.	Obs.	Exp.	Obs.	Exp.
Lip	45	3	3.5	20	21.5	23	25.0
Oesophagus	45	1	0.9	13	8.8	14	9.7
Stomach	30	9	10.7	63	54.8	72	65.5
Colon[a]	35	7	5.9	23	26.9	30	32.8
Rectum[a]	40	3	2.1	24	20.3	27	22.4
Pancreas	35	5	5.2	27	25.8	32	31.0
Larynx	50	2	4.6	29	27.4	31	32.0
Lung	30	35	37.0	327	369.7	362	406.7
Bladder[b]	40	7	3.4	52	30.2	59	33.6
Total[c]	.	72	73.3	578	585.4	650	658.7

[a] Villous adenomas excluded
[b] Papillomas excluded
[c] Sites listed above only

Table 4. Observed (Obs.) and predicted (Exp.) numbers of cancer cases in North Karelian females in 1982–86 at selected sites by age group

Site	Lowest age (LO)	Ages LO-54		55-74		Total	
		Obs.	Exp.	Obs.	Exp.	Obs.	Exp.
Lip	45	-	0.2	2	2.5	2	2.7
Oesophagus	45	-	0.7	14	13.7	14	14.4
Stomach	30	3	8.8	50	40.8	53	49.6
Colon[a]	35	3	5.2	39	30.7	42	35.9
Rectum[a]	40	3	3.9	18	25.8	21	29.7
Pancreas	35	1	3.0	30	28.5	31	31.5
Lung	30	5	3.8	25	22.9	30	26.7
Breast	30	81	91.4	119	131.4	200	222.8
Corpus uteri	35	13	14.5	47	36.9	60	51.4
Bladder[b]	40	-	0.8	12	6.8	12	7.6
Total[c]	.	109	132.3	356	340.0	465	472.3

[a] Villous adenomas excluded
[b] Papillomas excluded
[c] Sites listed above only

Figure 4. The observed and predicted larynx cancer cumulative incidence rates in males aged 55–74 years in North Karelia (NK), Kuopio county (KU) and other counties in Finland (OC) in 1957–86 by calendar period and the predicted rates (x) for North Karelia

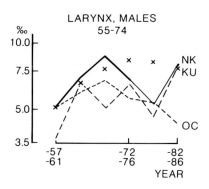

Figure 5. The observed and predicted stomach cancer cumulative incidence rates in males and females aged 30–54 and 55–74 years in North Karelia (NK), Kuopio county (KU) and other counties in Finland (OC) in 1957–86 by calendar period and the predicted rates (x) for North Karelia

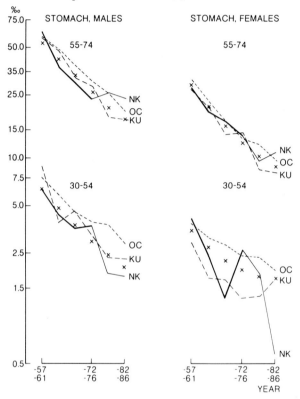

Intestines. The observed colon cancer incidence in North Karelian males was slightly lower than that predicted in ages 55–74 years in 1977–86 (Figure 6). In this period there was significant ($p < 0.05$) heterogeneity in the pattern of the incidence curves between the three areas. In contrast to the males, the observed number of colon cancer cases in females in North Karelia slightly exceeded the predicted number in 1982–86 (Tables 3 and 4). In cancer of the rectum, the situation was the reverse: higher than predicted numbers in males, lower than predicted numbers in females.

Pancreas. Despite the great increase in the observed incidence of pancreatic cancer in 1957–76 in North Karelian males, the observed and predicted incidence rates in 1982–86 in ages 55–74 years agreed fairly well (Figure 6, Table 3). The numbers also agreed well for females (Table 4).

Oesophagus and lip. The incidence rates of oesophageal cancer behaved in much the same way as the rates for stomach cancer in males (Figure 5). The rates for the North Karelian males were higher than those predicted in 1977–86 and

Figure 6. The observed and predicted colon cancer cumulative incidence rates for (a) colon cancer and (b) pancreas cancer in males aged 55-74 years in North Karelia (NK), Kuopio county (KU) and other counties in Finland (OC) in 1957-86 by calendar period and the predicted rates (x) for North Karelia

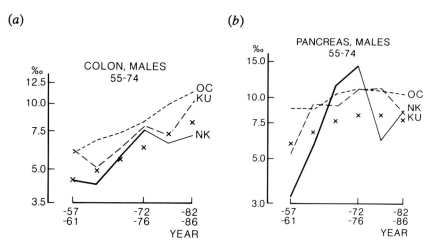

deviated significantly ($p < 0.05$) from the previous trend (see also Table 3). No such deviation was observed for females (Table 4). The observed numbers of lip cancer cases in 1982-86 were close to those predicted in both sexes.

Breast. The observed incidence of breast cancer in females in North Karelia was somewhat lower than that predicted in 1977-86 (Figure 7). However, none of the changes and differences were statistically significant. In terms of numbers of cases in 1982-86, this decrease was equivalent to a 10% reduction of cases in ages 30-54 and 55-74 years (Table 4).

Corpus uteri. Before 1977, the cumulative incidence rate in cancer of the corpus uteri in North Karelia had been consistently around 0.5% in ages 55-74, but it almost doubled in the following ten-year period, reaching the level for the whole country in 1977-86. There was significant ($p < 0.01$) heterogeneity in the pattern of the incidence curves in 1977-86 between the three areas. The observed number of cases in North Karelia exceeded that predicted for 1982-86 (Table 4).

Bladder. For both males and females of ages 55-74, the patterns of the incidence curves of cancer of the bladder were very similar in North Karelia and Kuopio county, but a steep rise in the incidence occurred five years earlier in Kuopio county. Before 1977, the rates both in North Karelia and in Kuopio county were lower than those in other counties, but, in 1982-86, the rates in these two counties were slightly higher than in the other counties. There was significant ($p < 0.01$) heterogeneity in the pattern of the incidence curves for males in 1977-86 between the three areas, and the course of the curve in North Karelia in those years differed highly significantly ($p < 0.001$) from that of the previous years. The observed number of cases of bladder cancer in North Karelia was much higher than that predicted in 1982-86 for both sexes (Tables 3 and 4).

Figure 7. The observed and predicted breast cancer cumulative incidence rates in females aged 30–54 and 55–74 years in North Karelia (NK), Kuopio county (KU) and other counties in Finland (OC) in 1957–86 by calendar period and the predicted rates (x) for North Karelia

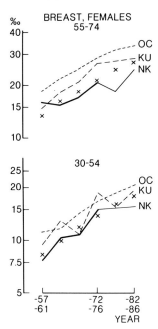

Discussion

The existence of reliable information on the trends both in certain risk factors and in the incidence of different cancers in the same population provides a good opportunity to study whether the risk-factor intervention in North Karelia influences the cancer-incidence rates.

Population surveys carried out in the two areas of eastern Finland since 1982 indicate that the CVD-related lifestyles and risk factors, especially smoking, some dietary habits and serum cholesterol levels, changed favourably during the community-based intervention in North Karelia. Changes in the same direction, although generally smaller, also occurred in the county of Kuopio, the reference area for the evaluation of the North Karelia project. In particular, the decrease in smoking among men was more marked in North Karelia than in Kuopio county. The changes in the risk factors observed in the county of Kuopio should reflect trends in the country as a whole. However, it is possible that the changes there have been greater than in the rest of Finland because of the extensive publicity given to the North Karelia project in the adjacent county and because large population surveys were also carried out in this area at five-year intervals.

In order to draw reliable conclusions, strictly comparable data on the development of risk factors should also be available for the rest of Finland and for North Karelia and the county of Kuopio for periods before the North Karelia

Project started. The intervention populations should also be larger in order to provide larger numbers of cancer cases.

It is obvious that there is a time lag before changes in risk factors will result in changes in the incidence rates of a disease and that this time lag is greater for cancer than for CVD. Therefore, it may be unlikely that any major impact of the risk-factor changes that started in 1982 would yet be apparent. However, in some instances, the effect of eliminating a risk factor (e.g., smoking) is seen rather soon. Hence, towards the end of the observation period some changes may have taken place. Because the rates given here are based on five-year periods, some small recent real changes may thus remain undetected.

The main finding regarding smoking-related cancers was the decline in the incidence of lung cancer among men. This decline started in about 1970 and was somewhat more marked in the two counties of eastern Finland where the rate in the 1960s was far above the national mean. The observed change in lung cancer incidence is in good agreement, allowing for some time lag, with the reduction in the prevalence of smoking among men since the early 1960s. In the 1960s, the prevalence of smoking among men was higher in eastern Finland than in the western part of the country, and the reduction started later in the east than in the west (Karvonen et al., 1967). The possible effect of the North Karelia project on the lung cancer incidence trend is largely hidden in the general decrease which started before the programme in eastern Finland.

The reduction in the incidence of lung cancer compared with the predicted rate was most marked for males aged 65–74. The main target group for the North Karelia project was males aged 25–59 in 1972. These people would be 37–71 years of age in the middle of the last five-year observation period, in 1984. The published figures (Puska et al., 1983) indicate that the reduction of smoking was greater among older people in this range. On the other hand, the observed and predicted incidence rates did not differ for laryngeal cancer in males in 1982–86.

The incidence of stomach cancer declined consistently and markedly in eastern Finland and in the rest of Finland. No differences were found in the trends between different areas before 1977. However, in 1977–86 an unexpected increase was observed in the numbers of cases compared with those predicted for both stomach and oesophageal cancer in males aged 55–74. A similar tendency was observed in stomach cancer in females. It is speculative whether this could be an adverse effect of the recommended increased consumption of cereals (Puska et al., 1981) in North Karelia.

There is a large body of evidence concerning the elevated risk of stomach cancer and high carbohydrate intake (Howson et al., 1986). Trends in oesophageal cancer in Finland have closely followed those for stomach cancer, and dietary factors have been assumed to be responsible for this (Hakulinen et al., 1986a).

The recommended dietary patterns in North Karelia could lead to a decreased risk of colon cancer, taking into consideration risk ratios larger than one, reported in many epidemiological studies for high consumption of meat and fat, and low consumption of vegetables and fibre (Jensen, 1985). On the other hand, the role of low cholesterol levels *per se* in the causation of colon and other cancers is still under speculation (McMichael et al., 1984; Knekt et al., 1988b). The numbers observed and predicted in North Karelia do not permit any definitive conclusions.

The North Karelia project placed great emphasis on decreasing consumption of fat and meat. In addition to colon cancer, a decrease in the risk of breast cancer might be expected (Rohan & Bain, 1987). A non-significant decrease compared with the predicted number was indeed observed in women aged 30–54 and 55–74. The coming years will show whether this finding is due to chance or whether the decline in the numbers of cases will continue. Fat has also been implicated in the etiology of corpus uteri cancer (Hill, 1985). The sudden rise in the incidence of that cancer in North Karelia in 1977–86 contradicts this suggestion and may be a consequence of other factors.

There is a fair amount of evidence for the possible general preventive role of a diet rich in vitamins A, C, and E for cancer at any site (Armstrong & Mann, 1985; Knekt et al., 1988a). With the data available, it is very difficult to assess any such tendency. The North Karelia project recommended increased use of foods containing these vitamins. The only general tendency for a lower than predicted risk was observed in cancers in North Karelian females under the age of 55.

The marked increase in bladder cancer incidence in North Karelia, compared with the expected incidence in 1977–86, is at first glance at variance with the known etiology of tobacco smoking and bladder cancer (IARC, 1986). However, the difficulty caused by the changing definition of bladder cancer for trend analyses and incidence predictions has been noted previously (Hakulinen et al., 1986a,b). Some of the lesions that would have been classified as papilloma in the 1950s are classified as carcinoma of the bladder in the 1980s and thus reported to the Cancer Registry as true malignancies. The temporal development of the incidence rates suggests that, on average, the change in classification has taken place later in Kuopio county than in other counties and in North Karelia about five years later than in Kuopio county.

Excluding cancer of the bladder, the numbers of both smoking-related and fat-related cancer cases in North Karelia in 1981–86 were smaller than predicted. The observed number of smoking-related cancer cases (cancers of the lip, oesophagus, pancreas, larynx and lung) was 539 and that predicted 579.7, the difference being entirely due to lung cancer in males. The number of fat-related cancer cases (cancers of the colon, rectum, breast and corpus uteri) was 323 and that predicted 339.8, the difference being attributable to breast cancer.

The statistical methodology employed may appear complicated or too sophisticated. Instead of predicting numbers of cases for North Karelia on the age-specific rates for the pre-intervention period by sex, or on the trends observed in the Kuopio county during and after the intervention, the present methodology made it possible to take into account trends and differences in trends observed during the periods before intervention both in North Karelia and elsewhere in Finland. The possible existence of high-risk cohorts before, during and after intervention could also adequately be taken into account or controlled for. The inclusion of the data from counties other than Kuopio as a reference made the comparison more powerful for individual cancers that are far less common than cardiovascular deaths. Moreover, the majority of other counties in Finland did not share a border with North Karelia and the effect of the North Karelia project on risk factors is likely to be less pronounced in these regions than in the neighbouring Kuopio county. On the other hand, the basic incomparability of the reference rates was slightly increased with this choice.

A derivation of the predicted numbers of lung cancer cases based on changes in smoking habits might have been possible (Hakulinen & Pukkala, 1981; Hakulinen et al., 1987). In this study, however, the main emphasis was not on showing how successfully future lung cancer incidence rates may be predicted using information on smoking. The aim was rather to study whether the incidence rates that could be predicted for North Karelia actually materialized in the absence of unexpected changes in smoking patterns.

The statistical methods used to predict the incidence of cancer in North Karelia in 1977–86 were based on the trends observed in all three areas involved and on the deviations from those trends by birth-year cohort and by period and on a common pattern of age-specific risks. First, fairly simple models were tried; more complicated ones were attempted only when the simpler model did not give a sufficient fit. A good fit is by no means a guarantee of a good prediction, especially when the model involves a large number of parameters.

Special problems are also caused by the inherent non-identifiability involved in the age–period–cohort models (Clayton & Schifflers, 1987). As Clayton and Schifflers suggest, the place of age–period–cohort models may well be restricted to descriptive epidemiology rather than forecasting. Actually, no real forecasting was involved in this study. With the exception of lung cancer in males, the future-period effects for North Karelia were estimated from data containing information on the prediction period, but obtained from other counties. The prediction interval was only ten years, which does not involve undue assumptions of continuity of the drift. For lung cancer in males, the model used in the prediction did not include any non-linearity with respect to period effects.

The choice of model had a surprisingly small effect on the predicted numbers of new cancer cases. Cancer of the corpus uteri is an extreme example. The model with power terms for cohort and period did not fit particularly well but gave a predicted number of cases, 37.2 for North Karelian females aged 55–74 in 1982–86. A better fit was achieved by applying a categorical period variable; the predicted number of cases was then 35.9. The fit of this model could be further improved by inserting a categorical cohort variable instead of the cohort power terms, and the predicted number was finally 36.9 (Table 4). For the other sites, changes in the models led to very small changes in the expected numbers. However, for lung cancer in males the model with categorical cohort and period variables for the three areas combined (Figure 1) gave an expected number of 416.3 for ages 30–74, which is not too far from the final predicted number of 406.8 (Table 3). When a full age–period interaction (including all cohort effects) was included in the model, the expected number was 417.5. Of course, the differences in the age groups were larger. For example, the model in Figure 1 predicted 49.7 lung cancer cases for ages 30–54, the model including the age–period interaction predicted 48.7, and the final model predicted 37.0 cases

(Table 3). In any case, it is to be noted that the predicted numbers are also random quantities.

The cumulative incidence rates used in Figures 3–7 are perhaps not the best indices for describing the cancer risk over the whole age range as they are dominated in general by the risk in the highest five-year age group. On the other hand, these quantities are not dependent on a standard population and have a concrete interpretation. They estimate the probability of a person alive at the beginning of the age range contracting the cancer in question before the end of the age range, in the absence of competing risks of death. For lung cancer in North Karelian males aged 55–74 that estimate is, for example, of the order of 10% (Figure 3). For pancreatic cancer the order of magnitude of this figure is 1% (Figure 6).

The trends in cancer of the lung in males and cancer of the breast in females in North Karelia are in general agreement with the changes in the proposed risk factors, that is, smoking and consumption of fat. It is speculative whether the findings for cancers of the stomach and oesophagus included adverse effects of the change in diet. The next ten years may show whether the changes in lifestyle and cancer risk are persistent in North Karelia and in the rest of the country.

Acknowledgement

This study was supported by the Cancer Institute of Finland.

References

Armstrong, B.K. & Mann, J.I. (1985) Diet. In: Vessey, M.P. & Gray, M., eds, *Cancer Risks and Prevention*, Oxford, Oxford University Press, pp. 68-98

Clayton, D. & Schifflers, E. (1987) Models for temporal variation in cancer rates. II. Age-period-cohort models. *Stat. Med.*, 6, 461-481

Hakulinen, T. & Pukkala, E. (1981) Future incidence of lung cancer: forecasts based on hypothetical changes in the smoking habits of males. *Int. J. Epidemiol.*, 10, 233-240

Hakulinen, T., Andersen, A., Malker, B., Pukkala, E., Schou, G. & Tulinius, H. (1986a) Trends in cancer incidence in the Nordic countries. A collaborative study of the five Nordic cancer registries. *Acta Pathol. Microbiol. Scand.*, Sect. A, Suppl. 288

Hakulinen, T., Teppo, L. & Saxén, E. (1986b) Do the predictions for cancer incidence come true? Experience from Finland. *Cancer*, 57, 2454-2458

Hakulinen, T., Magnus, K. & Tenkanen, L. (1987) Is smoking sufficient to explain the large difference in lung cancer incidence between Finland and Norway? *Scand. J. Soc. Med.*, 15, 3-10

Hill, M.J. (1985) Diet and human cancer: a new era for research. In: Joossens, J.V., Hill, M.J. & Geboers, J., eds, *Diet and Human Carcinogenesis*, Amsterdam, Excerpta Medica, pp. 3-12

Howson, C.P., Hiyama, T. & Wynder, E.L. (1986) The decline in gastric cancer: epidemiology of an unplanned triumph. *Epidemiol. Rev.*, 8, 1-27

IARC (1986) *IARC Monographs on the Evaluation of the Carcinogenic Risk of Chemicals to Humans*, Vol. 38, Tobacco Smoking, Lyon, International Agency for Research on Cancer

Jensen, O.M. (1985) The role of diet in colorectal cancer. In: Joossens, J.V., Hill, M.J. & Geboers, J., eds, *Diet and Human Carcinogenesis*, Amsterdam, Excerpta Medica, pp. 137-147

Karvonen, M., Blomqvist, G., Kallio, V., Orma, E., Punsar, S., Rautaharju, P., Takkunen, J. & Keys, A. (1967) Epidemiological studies related to coronary heart disease: characteristics of men aged 40-59 in seven countries. C4. Men in rural east and west Finland. *Acta Med. Scand.*, Suppl. 169

Knekt, P., Aromaa, A., Maatela, J., Aaran, R., Nikkari, T., Hakama, M., Hakulinen, T., Peto, R., Saxén, E. & Teppo. L. (1988a) Serum vitamin E and risk of cancer among Finnish men during a 10-year follow-up. *Am. J. Epidemiol.*, *127*, 28-41

Knekt, P., Reunanen, A., Aromaa, A., Heliövaara, M., Hakulinen, T. & Hakama, M. (1988b) Serum cholesterol and risk of cancer in a cohort of 39,000 men and women. *J. Clin. Epidemiol.*, *41*, 519-530

McMichael, A.J., Jensen, O.M., Parkin, D.M. & Zaridze, D.G. (1984) Dietary and endogenous cholesterol and human cancer. *Epidemiol. Rev.*, *6*, 192-216

Ovcarov, V. & Bystrova, V. (1978) Present trends in mortality in age group 35-64 in selected developed countries between 1950-73. *World Health Stat. Quart.*, *31*, 208-346

Payne, C.D., ed. (1986) *The Generalized Linear Interactive Modelling System Release 3.77 Manual*, Oxford, Numerical Algorithms Group

Puska, P., Tuomilehto, J., Salonen, J., Nissinen, A., Virtamo, J., Björkqvist, S., Koskela, K., Neittaanmäki, L., Takalo, T., Kottke, T.E., Mäki, J., Sipilä, P. & Varvikko, P. (1981) *Community Control of Cardiovascular Diseases. Evaluation of a Comprehensive Community Programme for Control of Cardiovascular Diseases in North Karelia, Finland 1972-1977*, Copenhagen, World Health Organization

Puska, P., Salonen, J., Nissinen, A., Tuomilehto, J., Vartiainen, E., Korhonen, H., Tanskanen, A., Rönnqvist, P., Koskela, K. & Huttunen, J. (1983) Change in risk factors for coronary heart disease during 10 years of a community intervention programme (North Karelia Project). *Br. Med. J.*, *287*, 1840-1844

Rohan, T.E. & Bain, C.J. (1987) Diet in the etiology of breast cancer. *Epidemiol. Rev.*, *9*, 120-145

Saxén, E. & Teppo, L. (1978) *Finnish Cancer Registry 1953-1978. Twenty-five Years of a Nationwide Cancer Registry*, Helsinki, Finnish Cancer Registry

Waterhouse, J., Muir, C., Shanmugaratnam, K., Powell, J., Peacham, D., Whelan, S. & Davis, W. (1982) *Cancer Incidence in Five Continents*, Vol. IV (IARC Scientific Publications No. 42), Lyon, International Agency for Research on Cancer

Evaluating Effectiveness of Primary Prevention of Cancer
Ed. M. Hakama, V. Beral, J.W. Cullen & D.M. Parkin
Lyon, International Agency for Research on Cancer
© IARC, 1990

A PRIMARY PREVENTION STUDY OF ORAL CANCER AMONG INDIAN VILLAGERS. EIGHT-YEAR FOLLOW-UP RESULTS

P.C. Gupta[1], F.S. Mehta[1], J.J. Pindborg[2]
D.K. Daftary[1], M.B. Aghi[1], R.B. Bhonsle[1] and P.R. Murti[1]

*[1]Basic Dental Research Unit,
Tata Institute of Fundamental Research,
Bombay, India
[2]Department of Oral Pathology,
Royal Dental College, Copenhagen, Denmark*

It is well established that the high incidence of oral cancer in India and in several other South-East Asian countries is caused by the use of tobacco (IARC, 1985). Most of this oral cancer – as much as 90% according to WHO estimates – is directly attributable to the chewing and smoking of tobacco (WHO, 1984). This indicates that oral cancer is amenable to primary prevention. The current study was undertaken to assess the feasibility and effectiveness of a primary prevention programme for oral cancer in a rural Indian population.

In large-scale house-to-house cross-sectional surveys of over 150 000 individuals in rural India, the habits of tobacco chewing and smoking were strongly associated with oral cancer and precancer (Mehta *et al.*, 1969, 1972). A ten-year follow-up study of 30 000 individuals in three areas showed that new cases of oral cancer and oral precancer developed exclusively among tobacco chewers and smokers, even though a large proportion of the cohort consisted of individuals without any tobacco habit. Also, all new oral cancers developed among individuals with a prior diagnosis of oral precancerous lesions (Gupta *et al.*, 1980). When tobacco use was stopped or reduced substantially, the regression rates of oral precancerous lesions increased significantly (Mehta *et al.*, 1982). These results established an almost complete association between tobacco use and oral cancer and precancer.

The present study was undertaken with two objectives: (i) to find out whether rural individuals can be motivated to give up their tobacco habits through a concentrated programme of health education and (ii) whether such a programme of health education would affect the incidence or the risk of oral cancer. This on-going study is being conducted in three districts of India, and the eight-year follow-up results from one area, Ernakulam district in Kerala state in southern

India (where the incidence of oral cancer and precancerous lesions is known to be high) are discussed.

Materials and methods

Two distinct cohorts were selected and followed up at regular intervals in Ernakulam district. In both cohorts the baseline and the annual follow-up consisted of an interview and a clinical mouth examination of each individual in house-to-house surveys. In follow-up surveys each individual was identified before the interview and examination, with the help of an alphabetical list of names, cross-indexed with addresses and other identifying information.

The most common method of tobacco smoking was bidi smoking. Bidi is a cheap smoking stick 4–8 cm in length, consisting of a rolled piece of dried temburni leaf (*Diospyros melanoxylon*) with 0.15 to 0.25 g of coarse ground tobacco. Tobacco was chewed most commonly in the form of a betel quid consisting of betel leaf, lime (calcium hydroxide), areca nut, and tobacco. More details regarding tobacco habits are given elsewhere (Mehta et al., 1969).

The most common oral precancerous lesion was oral leukoplakia. Leukoplakia was defined as a raised white patch 5 mm or more in diameter which could not be scraped off and could not be attributed to any other disease. This definition carried no histological connotation. Following the study protocol, no surgical or medicinal treatment was recommended for oral leukoplakia. If leukoplakia (or any other lesion in the mouth) was judged as suspicious by the examining dentist, a biopsy was done with the patient's consent. Microscopically, if the biopsy showed carcinoma, the patient was counted as an oral cancer case, referred for treatment, and was withdrawn from the study.

The survey teams that went house-to-house consisted of dentists, interviewing clerks, drivers and local helpers. For the intervention cohort a social scientist was also a member of the team. The intervention cohort was subjected to a concentrated programme of health education concerning tobacco use in various different forms, whereas the control cohort was not subjected to any such campaign.

Intervention cohort

A complete screening of the population from selected *karas* (the smallest population unit available through census publications) in the Ernakulam district was carried out and all available tobacco users aged 15 years and over were chosen (12 212 individuals) as the study sample. Only temporary residents, the very old, sick, infirm, psychologically disturbed, and those with treated oral cancers were excluded. The baseline survey for the intervention cohort was conducted in 1977–78 and the present report covers eight annual follow-ups thereafter.

Special studies revealed that many people began to use tobacco and very often continued use because of its perceived medicinal value for disorders such as toothache and gastric disturbances. There was almost no awareness of any possible adverse health consequence of tobacco use. The health education programme therefore consisted of two broad categories: (1) information for creating awareness regarding the relationship between tobacco and oral cancer and for convincing a target population of this relationship, and (2) helping individuals with strategies for stopping their tobacco use. In the health education campaign, personal as well as mass-media communication was employed. The personal communication was given by the dentist after oral examination and then by the social scientist according to a pre-designed format. This personal communication was given once a year at the time of follow-up examination.

To create an awareness regarding tobacco habits, a special documentary film was produced and screened for the target population. Printed pictorial posters were placed in the villages, and messages were broadcast by radio. Special articles were written in newspapers.

After some time when the population became aware, and more or less convinced, of the relationship between tobacco habits and oral cancer, the emphasis of the health education was gradually shifted to the explanation of possible strategies for stopping tobacco use. For this purpose one more documentary film was made. Individualized, handwritten posters were placed in each village. The personal communication continued to be a major and important health education resource.

The health education programme was on-going, dynamic, and responsive to the feedback from the target population. All the intervention inputs were pre-tested before implementation and modified if necessary. The entire health education material was solely based upon scientific facts.

Control cohort

For this cohort, *karas* were selected by random sampling and their entire population aged 15 years and above was examined. The baseline survey was conducted in 1966–67 and the first follow-up survey three years later. A total of eight annual follow-up surveys were conducted providing ten-year follow-up results. For the purpose of this report, only a subset of the original cohort was included – namely tobacco users in the baseline survey and the first eight-year results of the ten-year follow-up study.

Although health education was not actively attempted in the control cohort, the conduct of the study itself provided some intervention inputs. The association between oral cancer and tobacco habits was explained to patients, especially while interviewing and conducting oral examinations. The examining dentists routinely advised patients to give up their tobacco habits, more forcefully if the patient had an oral precancerous lesion.

Statistical analysis

The incidence rates of leukoplakia were calculated using person-year methods. The numerator for incidence rate consisted of individuals with a diagnosis of oral leukoplakia in any subsequent follow-up who did not have a diagnosis of oral leukoplakia, submucous fibrosis or oral cancer in any of the previous examinations. The denominator consisted of person-years of observation among individuals without any prior diagnosis of oral leukoplakia, submucous fibrosis or oral cancer. The age and tobacco use varied over time and these were taken into account. The expected number of leukoplakias in the intervention cohort was calculated using age- and sex-specific incidence rates from the control cohort.

The changes in tobacco habits were classified in four categories. Stopping the use of tobacco was defined as complete discontinuation of any form of tobacco use at least six months before the interview. Reduction was defined as 50% or more decrease in the frequency of tobacco use for those practising one single tobacco habit. For those practising multiple tobacco habits of chewing and smoking, reduction was defined as at least 50% reduction in the frequency of each of the tobacco habits or complete stoppage of one tobacco habit without increasing the frequency of other tobacco habits. The increased category was defined as 50% or more increase in the frequency of tobacco use. All other individuals were put in the unchanged category.

Results

In the baseline survey the intervention cohort consisted of 12 212 tobacco users and the control cohort of 6067 tobacco users. The loss to follow-up was 0.8% in the intervention cohort and 10.3% in the control cohort. The annual follow-up percentage varied from 93.2% to 83.6% in the intervention cohort and 74.2% to 70.2% in the control cohort.

Table 1 shows changes in tobacco habits in the cohorts over the eight-year period among men and women. The most remarkable differences between the intervention and control cohorts are observed in the 'stopped' and the 'increased' categories. Among men, the percentage of individuals who stopped their tobacco use was 2.2% in the control cohort compared with 9.5% in the intervention cohort and among women it was 6.1% compared with 19%. The percentage of men increasing their tobacco use was 30.8% in the control cohort, compared with 11.4% in the intervention cohort, and among women the percentages were 14.1% and 2.2%. The percentage who reduced their tobacco habit was higher in the intervention cohort than the control cohort. Thus the effect of intervention was visible over the entire range of change in tobacco use.

Table 1. A comparison of changes in tobacco habits among intervention and control cohort after eight years of follow-up

	Percentage change			
	Men		Women	
	Intervention	Control	Intervention	Control
Increased	11.4	30.8	2.3	14.1
Unchanged	48.3	47.0	41.1	46.4
Reduced	30.8	20.0	37.6	33.4
Stopped	9.5	2.2	19.0	6.1
No. of persons	7141	2858	3069	1564

Table 2 shows changes in tobacco use for specific tobacco habits among men and Table 3 shows similar results for women. Among men in the category of 'stopped', the most remarkable difference was seen in chewers, where percentage giving up the habit was 1% in the control cohort compared with 13% in the intervention cohort. In the 'increased' as well as 'reduced' categories, the most prominent differences were seen among the bidi smokers. In both cohorts, almost all women were tobacco chewers and the differences in both categories, 'stopped' as well as 'increased', were substantial.

Table 2. Changes in tobacco use after eight years of follow-up among men

	Percentage change							
	Bidi		Cigarette		Chewers		Mixed	
	Intervention	Control	Intervention	Control	Intervention	Control	Intervention	Control
Increased	13	41	32	54	9	26	3	10
Unchanged	51	47	36	27	43	42	47	54
Reduced	27	10	9	7	35	31	45	35
Stopped	9	2	23	12	13	1	5	1
No. of persons	4583	1432	250	163	927	470	1381	793

Table 3. Changes in tobacco use after eight year of follow-up among women

	Percentage change					
	Bidi		Chewing		Mixed	
	Intervention	Control	Intervention	Control	Intervention	Control
Increased	10	22	2	14	–	9
Unchanged	40	33	42	47	25	18
Reduced	10	17	39	33	54	69
Stopped	40	28	18	6	21	9
No. of persons	145	18	2856	1535	68	11

Table 4 shows the annual age-adjusted incidence rate (per 1000) after eight years of follow-up. It is clear that the incidence of leukoplakia was significantly and substantially lower in the intervention cohort than in the control cohort, for men as well as women, and in all categories of tobacco users.

Table 5 shows the observed and expected numbers of leukoplakias in the intervention group for men and women in each category of tobacco use. The expected numbers of leukoplakias was calculated on the basis of age-specific incidence rates in the control cohort separately for each cell in the table. Among men, the largest proportionate decrease occurred among chewers, but the largest absolute decrease occurred among smokers and men with mixed habits. Overall, the observed leukoplakias in men were only 41% of the expected number, and in women only 28% of the expected number.

Table 4. Annual age-adjusted incidence rate (per 1000) of leukoplakia after eight years of follow-up

Tobacco habit	Leukoplakia incidence			
	Men		Women	
	Intervention	Control	Intervention	Control
Bidi smoking	1.7	4.0	–	9.0
Cigarette	–	3.9	–	–
Chewing	3.5	7.4	4.8	6.2
Mixed	4.1	11.1	2.7	4.4
Overall	2.5	6.8	3.5	6.2

Table 5. Observed and expected numbers of leukoplakias in the intervention group after eight years of follow-up

Tobacco habit	Numbers of leukoplakia cases					
	Men			Women		
	Observed	Expected	Ratio	Observed	Expected	Ratio
Smoking						
Bidi	63	142.6	0.44	0	12.8	0.0
Cigarette	0	2.6	0.0	0	0	–
Chewing	27	79.7	0.34	49	163.9	0.30
Mixed	59	136.1	0.43	2	3.8	0.53
Overall	149	361.0	0.41	51	180.5	0.28

Table 6. Age distribution of observed and expected numbers of leukoplakias in intervention group after eight years of follow-up

Age group	Numbers of leukoplakia cases					
	Men			Women		
	Observed	Expected	Ratio	Observed	Expected	Ratio
15–34	11	21.4	0.51	3	7.5	0.40
35–54	84	188.3	0.45	15	107.9	0.14
55+	54	151.2	0.36	33	65.1	0.51
Overall	149	360.9	0.41	51	180.5	0.28

In the intervention cohort, 19 new oral cancers were detected over the eight-year follow-up period. For 18 of these, the diagnosis was confirmed histologically as squamous cell carcinoma. One patient refused biopsy but the course of disease and the associated clinical features made the diagnosis imperative. Based on the incidence rate of oral cancer in the control cohort (Gupta et al., 1980), 20.4 oral cancers were expected.

Discussion

These results corroborate the findings reported earlier after five years of follow-up (Gupta et al., 1986a,c), some of them in a more convincing manner. For example, a conclusion that, for stopping tobacco use, the health education was more helpful to male chewers than to other tobacco users was reached only after multiple logistic regression analyses of the data (Gupta et al., 1986a,b). This fact is now very apparent, as the percentage of male chewers who stopped their tobacco use in the intervention cohort is 13 times the percentage in the control cohort. This increase was much higher than for any other category of tobacco use.

Several new findings also emerge from this analysis. For example, it is clear that a major effect of the health education has been to discourage people from increasing the use of tobacco. This is interesting because the aim of the health education was only to encourage people to stop tobacco use. In view of a positive dose–response relationship between tobacco use and oral leukoplakia (Gupta, 1984), it can be hypothesized that this decrease in the percentage of those increasing the frequency of tobacco use has also contributed, perhaps substantially, to a reduction in the risk of leukoplakia.

The relationship between age and change in tobacco habit is interesting; it corroborates an earlier result based on five-year follow-up data that the effect of health education increased with increasing age. For women, however, it is not clear why there was such a sudden and substantial drop in the risk for the middle age group.

It appears that the eight-year follow-up period was not sufficient to show a significant decrease in the risk of oral cancer. It has, however, already been demonstrated that leukoplakia is the most important oral precancerous lesion in terms of being most common as well as being the point of origin for most oral cancers (Gupta et al., 1980). Therefore a decrease in the risk of leukoplakia can be construed as demonstrating the decrease in the risk of oral cancer. Thus, this study shows that primary prevention of oral cancer is a feasible and practicable proposition even among populations who have many misconceptions about supposed beneficial effects of tobacco.

This study does have some limitations. The most important limitation, perhaps, is the non-concurrence of the intervention and control cohorts. In terms of calendar time, the difference between the intervention and control cohort was ten years. Therefore the possibility that the observed differences could be due to differences in the time trends cannot be entirely ruled out. One method of assessing such a possibility is to compare the prevalence of leukoplakia in the baseline surveys of control and intervention cohorts. In the control cohort the prevalence was 2.7% among 6077 tobacco users (Mehta et al., 1969), and in the intervention control cohort it was 2.9% among 12 212 tobacco users (Mehta *et al.*, 1982). The difference was not significant (chi square - 0.7).

Although the possible benefits of health education in this study have been assessed only in terms of decrease in the risk of oral cancer, it should not be forgotten that tobacco use is responsible for increase in the risk of many diseases such as other cancers, especially those occurring in the aero-digestive tract, heart disease and respiratory diseases. It has been estimated that in India tobacco use is responsible for at least 635 000 extra deaths every year (Gupta, 1988). Thus, undoubtedly, the positive effects of health education could be far more than simply a decrease in the risk for oral cancer.

References

Gupta, P.C. (1984) A study of dose-response relationship between tobacco habits and oral leukoplakia. *Br. J. Cancer*, 50, 527-531

Gupta, P.C. (1988) Health consequences of tobacco use in India. *World Smoking & Hlth*, 13, 5-10

Gupta, P.C., Mehta, F.S., Daftary, D.K., Pindborg, J.J., Bhonsle, R.B., Jalnawalla, P.N., Sinor, P.N., Pitkar, V.K., Murti, P.R., Irani, R.R., Shah, H.T., Kadam, P.M., Iyer, K.S.S., Iyer, H.M., Hedge, A.K., Chandrashekhar, G.K., Shroff, B.C., Sahiar, B.E. & Mehta, M.N. (1980) Incidence of oral cancer and natural history of oral precancerous lesions in a 10-year follow-up study of Indian villagers. *Commun. Dent. Oral Epidemiol.*, 8, 287-333

Gupta, P.C., Aghi, M.B., Bhonsle, R.B., Murti, P.R., Mehta, F.S., Mehta, C.R. & Pindborg, J.J. (1986a) Intervention study of chewing and smoking habits for primary prevention of oral cancer among 12 212 Indian villagers. In: Zaridze, D.G. & Peto, R., eds, *Tobacco: A Major International Health Hazard* (IARC Scientific Publications No. 74), Lyon, International Agency for Research on Cancer, pp. 307-318

Gupta, P.C., Mehta, C.R., Pindborg, J.J., Aghi, M.B., Mehta, F.S., Bhonsle, R.B. & Murti, P.R. (1986b) Intervention of tobacco chewing and smoking habits. *Am. J. Public Hlth*, 76, 709

Gupta, P.C., Mehta, F.S., Pindborg, J.J., Aghi, M.B., Bhonsle, R.B., Murti, P.R., Daftary, D.K., Shah, H.T. & Sinor, P.N. (1986c) Intervention study for primary prevention of oral cancer among 36 000 Indian tobacco users. *Lancet*, i, 1235-1238

IARC (1985) *IARC Monographs on the Evaluation of the Carcinogenic Risk of Chemicals to Humans*, Vol. 37, *Tobacco Habits Other than Smoking; Betel-quid and Areca-nut Chewing; and Some Related Nitrosamines*, Lyon, International Agency for Research on Cancer

Mehta, F.S., Pindborg, J.J., Gupta, P.C. & Daftary, D.K. (1969) Epidemiologic and histologic study of oral cancer and leukoplakia among 50 915 villagers in India. *Cancer*, 24, 832-849

Mehta, F.S., Gupta, P.C., Daftary, D.K., Pindborg, J.J. & Choksi, S.K. (1972) An epidemiologic study of oral cancer and precancerous conditions among 101 761 villagers in Maharashtra, India. *Int. J. Cancer*, 10, 134-141

Mehta, F.S., Aghi, M.B., Gupta, P.C., Pindborg, J.J., Bhonsle, R.B., Jalnawalla, P.N. & Sinor, P.N. (1982) An intervention study of oral cancer and precancer in rural Indian populations: A preliminary report. *Bull. World Hlth Org.*, 60, 441-446

World Health Organization (1984) Control of oral cancer in developing countries. Report of a WHO meeting. *Bull. World Hlth Org.*, 62, 817-830

Evaluating Effectiveness of Primary Prevention of Cancer
Ed. M. Hakama, V. Beral, J.W. Cullen & D.M. Parkin
Lyon, International Agency for Research on Cancer
© IARC, 1990

PRIMARY PREVENTION OF CANCER: RELEVANT MULTIPLE RISK FACTOR INTERVENTION TRIAL RESULTS

W.T. Friedewald[1], L.H. Kuller[2] and J.K. Ockene[3]

[1]*Office of Disease Prevention, Office of the Director, National Institutes of Health, Bethesda, MD, USA*
[2]*Graduate School of Public Health, University of Pittsburgh, Pittsburgh, PA, USA*
[3]*Department of Medicine, Division of Preventive and Behavioural Medicine, University of Massachusetts Medical School, Worcester, MA, USA*

Introduction

The Multiple Risk Factor Intervention Trial (MRFIT) was a randomized, primary prevention clinical trial designed to test the effect of a multifactorial intervention programme on the mortality rate from coronary heart disease (CHD). The MRFIT screened 361 662 men, aged 35 to 57 years, during 1973–76 at 22 participating clinical centres in the USA. A total of 12 866 men without evidence of definite CHD by history, physical examination or electrocardiography (ECG) at rest were enrolled in the trial. The eligibility criteria, statistical design considerations and methods have been described in earlier publications (MRFIT Research Group, 1977, 1982). The selection criteria included the use of a multiple logistic risk function, derived from the Framingham Heart Study data, to estimate each prospective participant's risk of future CHD on the basis of his serum cholesterol concentration, diastolic blood pressure and cigarette smoking status, the three major modifiable CHD risk factors. Men selected at screening were those in the upper 15% (later 10%) of the risk score distribution. During screening, men with evidence of CHD, serious life-threatening conditions, markedly elevated diastolic blood pressure (≥ 115 mm Hg), or very high serum cholesterol levels (≥ 350 mg/dl) were excluded from the trial.

MRFIT participants were randomly assigned to either of two groups: a special-intervention programme (the SI group) or a group of participants who were referred to their customary sources of health care in the community (the usual-care or

UC group). Both groups were systematically followed from randomization until 29 February 1982 – an average of 6.9 years. Following this formal trial period of active intervention and/or surveillance, the clinical participants were followed for vital status using central data sources until 31 December 1985 (MRFIT Group, 1990). Similarly, the original 361 662 subjects screened were followed using central data sources for up to six years after screening (Sherwin et al., 1987). Thus, this report has three sources of data: (1) risk factor, morbidity and mortality data on the 12 866 randomized participants in the MRFIT until 19 February 1982, (2) mortality data from 29 February 1982 to 31 December 1985 on the 12 866 randomized MRFIT participants, and (3) mortality data for up to six years on the 361 662 MRFIT participants screened.

Intervention programme

The UC men continued to be followed by their usual source of medical care, had no intervention programme offered, but were invited to return once a year for a medical history, physical examination and laboratory studies. The results of the screening and annual examinations were given to their personal doctors who were informed of the scientific objectives of the study.

The detailed components of the SI programme have been published elsewhere (Benfari, 1981; Caggiula et al., 1981; Hughes et al., 1981; Cohen et al., 1981) and are summarized here. The initial phase of intervention was an intensive integrated effort to lower the three major risk factors. Immediately after randomization to the SI group, each cigarette smoker was advised individually by a study doctor of the need to stop smoking. Shortly thereafter, each SI man was invited with his spouse or friend to a series of weekly group discussions addressing all three risk factors; uniformity of structure and content was sought by the use of common protocols and educational material. Each discussion group included about ten men and met for about ten sessions.

After the initial intensive intervention phase, individual counselling (planned and executed by an intervention team, usually headed by a behavioural scientist and including nutritionists, nurses, doctors and general health counsellors), became the general approach for all three areas of intervention. Participants in the SI group were seen every four months, or more often as needed for intervention purposes. Every SI participant was monitored to assess changes in risk factor status, the ultimate objective being to reach the specific goals established for each individual.

Hypertension

Hypertension was considered present if the man reported having had anti-hypertensive medication prescribed for him by his personal doctor (regardless of blood pressure (BP) level), or if an untreated man was found to have a diastolic BP of at least 90 mm Hg on two consecutive monthly visits during the trial. Before drug prescription, weight reduction was attempted for overweight men. Drugs were prescribed according to a carefully staged protocol beginning with the use of either hydrochlorothiazide or chlorthalidone. Reserpine, hydralazine, guanethidine or certain other drugs were sequentially added if the BP goal had not been achieved. The protocol also included a provision for mild sodium restriction. Participants in the SI group who had been treated with BP medication by non-study doctors were usually transferred, with the permission of their private doctors, to the care of an MRFIT clinician.

Nutrition

The nutrition intervention programme sought to encourage the development of lifelong shopping, cooking and eating patterns rather than to specify a structural diet. Individual intervention goals to lower serum cholesterol levels by an amount dependent on the entry level were established. Serum cholesterol concentration was determined using automated methods with periodic quality control. Twenty-four hour dietary records were obtained through interview by an MRFIT nutritionist, and were coded by a Nutrition Coding Center (NCC) (Dennis et al., 1980).

Initially, eating patterns were recommended that reduced saturated fat intake to less than 10% of calories and dietary cholesterol intake to less than 300 mg/day, and increased polyunsaturated fat intake to 10% of calories. In 1976, the nutrition pattern was changed to specify that saturated fat be less than 8% of calories and dietary cholesterol less than 250 mg/day. Weight reduction was sought for men whose weight was 115% or greater of desirable weight by recommending reductions in caloric intake and increases in moderate forms of physical activity.

Smoking

Cigarette smoking was reported in two ways: (1) self-reported smoking behaviour based on a participant interview; and (2) thiocyanate-adjusted cessation rates using serum thiocyanate level as an objective measure of smoking behaviour to adjust self-reported cases. The smoking intervention programme for SI participants used conventional behavioural modification techniques throughout the trial; aversive techniques and hypnosis were used in selected instances during the final years. Particularly successful intervention approaches were the ten-week group sessions at the beginning of the trial and the five-day stop-smoking clinics during the final years. No systematic effort was made to alter the smoking habits of persons who smoked only pipes and cigars. Dosage reduction, including changing to cigarettes low in tar and nicotine, was recommended only as an intermediate step to cessation.

Data collection methods

On or about each anniversary of randomization, participants in both the SI and UC groups returned for assessment of risk-factor levels and morbidity status. Data collected at these annual visits, at screening, intervention, and four-monthly visits were sent to the coordinating centre for processing and analysis.

Mortality ascertainment

All participants were followed in the trial for a minimum of six years, with an average period of observation of approximately seven years. During the trial and at termination of active intervention on 28 February 1982, the vital status of each man was checked by the clinical centre staff using a variety of procedures. Vital status for the randomized cohort after 28 February 1982 and the original 361 662 participants was determined using the National Death Index and information provided by the Social Security Administration. Mortality ascertainment is estimated to be essentially 100% complete using these data sources and previously reported matching algorithms.

Cause-specific death rates for the participants and for the randomized cohort for the post-trial period from 28 February 1982 to 31 December 1985 are based on coding of death certificates by trained nosologists, according to the Ninth Revision of the International Classification of Diseases (ICD-9). Each death certificate was independently coded by two nosologists and disagreements were adjudicated by a third nosologist.

In published papers describing cause-specific mortality during the 6.9 years of the trial through 28 February 1982, cause of death was determined by a Mortality Review Committee which considered hospital records, autopsy reports and interviews with witnesses of the death as well as death certificate information. Information other than that on the death certificate was not routinely available for deaths which occurred after 28 February 1982. Cause-of-death classifications by the Committee and by the nosologists were made without knowledge of group assignment (SI or UC).

To determine survival status from 28 February 1982 (six years after the last day of randomization), a telephone or mail contact was attempted with each man not previously known to be deceased. The status of men not located by this procedure was sought using the files of the Social Security Administration and the services of a commercial firm specializing in methods of follow-up. Cause of death was assigned by a Mortality Review Committee, a three-member panel of cardiologists not associated with any MRFIT centre or the interim trial results. This committee, without knowledge of study group membership of the deceased, reviewed clinical records, hospital records, next-of-kin interviews, death certificates and reports of autopsies performed (31% of SI decedents, 33% of UC).

Results

Risk factor changes in the randomized cohort

The randomization process established two comparable groups at baseline as reported earlier (MRFIT Research Group, 1982; Sherwin et al., 1981). A necessary goal of the trial was to obtain adequate reductions, through intervention, of the three major modifiable CHD risk factors (Figure 1). For each of these, highly statistically significant ($p < 0.01$) differences between the SI and UC groups were observed at each annual visit.

By 12 months, average diastolic BP reductions from entry of 6.3 mm Hg for SI men and 2.5 mm Hg for UC men were observed (Table 1). By 72 months, these reductions were 10.5 mm Hg and 7.3 mm Hg, respectively. Of the men randomized, 19% reported at entry being prescribed anti-hypertension medication; at six years, 58% of SI men and 47% of UC men reported such prescription. The SI–UC difference in diastolic BP averaged over annual visits was 4%.

Figure 1. Mean risk factor levels by year of follow-up for Multiple Risk Factor Intervention Trial Research Group participants. SI, special intervention group; UC, usual care group; S, first screening visit

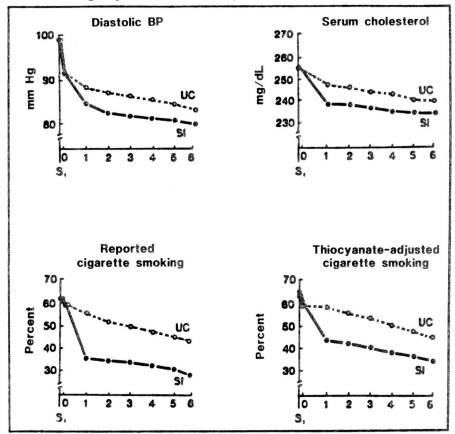

At the time of randomization, 59% of all men reported themselves as current cigarette smokers (Table 1). For men who reported smoking on entry, reported rates of those who stopped smoking at 12 months were 43% for SI men and 14% for UC men; at 72 months, these were 50% and 29%. Thiocyanate-adjusted rates of those who stopped smoking at 12 months were 31% for SI men and 12% for UC; at 72 months, these were 46% and 29%, respectively. At 72 months, the SI–UC differences in thiocyanate–adjusted rates of those who stopped smoking for the 22 centres ranged from 5% to 24%.

Table 1. Mean risk factor levels at entry and annual visits for MRFIT[a] special intervention group (SI) and usual care group (UC) men

	Baseline	Annual visits (months)					
		12	24	36	48	60	72
Diastolic BP (mm Hg)							
SI	91.0	84.7	82.5	82.0	81.6	81.2	80.5
UC	90.9	88.4	86.9	86.3	85.6	84.6	83.6
Reported cigarette smoking (%)							
SI	59.3	35.9	35.2	35.1	33.9	32.6	32.3
UC	59.0	55.6	52.2	50.5	48.2	46.7	45.6
Plasma cholesterol (mg/dl)							
SI	240.3	–	229.9	228.1	227.2	226.6	228.2
UC	240.6	–	237.2	235.1	234.7	232.3	233.1
No. of participants at each visit							
SI	6428	6112	5995	5883	5791	5662	5754
UC	6438	6080	5919	5793	5711	5615	5639

[a] MRFIT, Multiple Risk Factor Intervention Trial Research Group

Mean plasma cholesterol level at entry was 240 mg/dl. After two years, there were reductions of 10.4 mg/dl for SI men and 3.4 mg/dl for UC men; after six years the mean levels were 12.1 mg/dl and 7.5 mg/dl below entry levels for SI and UC men, respectively. These reductions after six years amount to an SI–UC difference in total cholesterol of 4.6 mg/dl, or 2%.

Data are not available on risk factor findings, treatments and non-fatal morbid events for the MRFIT cohort during the post-trial period from 28 February 1982 to 31 December 1985. Only mortality data are available for this period.

Mortality (cohort to 28 February 1982)

As of 28 February 1982, after an average period of follow-up of seven years, there were 260 deaths among UC men, of which 124 were ascribed to CHD and 145 to cardiovascular causes (including CHD). Of 265 SI deaths, 115 were ascribed to CHD and 138 to cardiovascular disease (CVD). The key mortality end-points of CHD and CVD were 7.1% and 4.7% less, respectively, in the SI compared with the UC group, while the death rate for all causes was 2.1% higher for the SI men. None of these differences is statistically significant.

The number of deaths from non-cardiovascular causes was also similar in the two groups (116 SI; 109 UC). There were 81 cancer deaths in the SI group and 69 in the UC, resulting from lung cancer (34 SI; 28 UC), colorectal cancer (8 SI; 6 UC), other gastrointestinal neoplasms (20 SI; 11 UC), and other neoplasia (19 SI vs 24 UC).

The number of deaths in the UC group was substantially less than expected for the six complete years of follow-up as well as for the average follow-up period of seven years. Based on assumption about changes in risk factors and the Framingham Heart Study risk functions, 442 deaths (including 187 from CHD) were expected by the end of six years of follow-up among the 6438 UC men; only 219 (including 104 from CHD) occurred. By the end of follow-up (an average of seven years) for all men, the total of 260 UC deaths (including 124 from CHD) was still well below the number expected for the six-year follow-up period.

Mortality (cohort to 31 December 1985)

Mortality follow-up until 31 December 1985 resulted in an average follow-up of 10.5 years with total follow-up of 134 954 person-years (67 470 for SI and 67 471 for UC).

CHD deaths numbered 202 in the SI group (3.1% of those originally randomized) and 266 (3.5%) in the UC group (Table 2), a relative difference of 10.6% ($p = 0.12$). CVD mortality was 8.3% lower for SI than UC ($p = 0.16$). Deaths from all causes numbered 496 in the SI and 537 in the UC group, a difference of 7.7% ($p = 0.10$). For other ischemic heart disease (ICD-9 411–414 and 429.2), rates were 11.7% higher for the SI than the UC group. Mortality from non-cardiovascular diseases was 7.1% lower for the SI than the UC group.

Table 2. MRFIT mortality end-points up until 31 December 1985 by treatment group

	SI ($n=6428$)		UC ($n=6438$)		Percentage difference	90% Confidence interval	p-value
	No.	%	No.	%			
CHD death[a]	202	3.1	226	3.5	−10.6	−23.7 to 4.9	0.12
CVD death[b]	266	4.1	290	4.5	−8.3	−20.2 to 5.5	0.16
All causes	496	7.7	537	8.3	−7.7	−16.6 to 2.3	0.10

[a] ICD-9 (410–414, 429.2)

[b] ICD-9 (390–459)

MRFIT, Multiple Risk Factor Intervention Trial

SI, special intervention group; UC, usual care group

Mortality data for cancer causes are presented in Table 3. Overall there was a 6.2% lower mortality in the SI group ('observed') compared to the UC group ('expected'). The largest percentage difference in favour of the SI group was for colorectal cancer (ICD-9, 153, 154). However, for lung cancer mortality, the SI group had a relative 20.5% higher rate (1.01% compared to 0.84%). None of these differences is statistically significant.

Table 3. Number and percentage of cancer deaths by primary site for MRFIT special intervention group (SI) and usual care group (UC) until 31 December 1985

Primary site	ICD-9 code	SI	UC	Percentage difference
Oral etc.	(140–149)	4 (0.06)	6 (0.09)	–
Digestive	(150–159)	34 (0.53)	42 (0.65)	−19.3
Colon/rectum	(153/154)	10 (0.16)	13 (0.20)	−22.8
Other	(150–152, 155–159)	24 (0.37)	29 (0.45)	−17.7
Respiratory etc.	(160–165)	66 (1.03)	55 (0.85)	+20.1
Lung	(162)	65 (1.01)	54 (0.84)	+20.5
All other		36 (0.56)	46 (0.71)	−22.0
Total cancer		140 (2.2)	149 (2.3)	−6.2

MRFIT, Multiple Risk Factor Intervention Trial

Results among cigarette smokers

The association between cigarette smoking and CHD and various forms of cancer has been documented from many data sources. Observational data from the MRFIT participants and cohort confirm these earlier multiple observations, most notably for cancer.

Mortality data by smoking history for the MRFIT participants

Six-year cause-specific mortality data for the 361 662 MRFIT participants categorized by cigarette smoking status are presented in Table 4. These are observational data and clearly different from the data from the randomized MRFIT cohort. Also there is no information on changes in cigarette smoking or other risk factors during the six-year period of mortality follow-up.

Cigarette smoking can be seen to be a risk factor for cancer of the lung, pancreas, oral cavity and larynx, oesophagus, kidney, bladder, and for coronary heart disease, stroke and total mortality.

Table 4. MRFIT participants – six-year follow-up. Cause-specific mortality by smoking levels

Cigarettes /day	Number screened	Lung cancer deaths	Age-adjusted /10 000	Pancreas cancer deaths	Age-adjusted /10 000	Mouth/larynx cancer deaths	Age-adjusted /10 000	Oesophagus cancer deaths	Age-adjusted /10 000	Kidney cancer deaths	Age-adjusted /10 000
0	228 545	218	9.18	69	2.89	10	0.43	28	1.14	42	1.76
1–19	29 333	60	21.66	6	2.12	5	1.78	9	3.22	3	1.05
20–39	72 200	323	47.74	34	4.94	21	3.16	23	3.27	34	5.04
40+	31 584	230	78.62	20	6.82	21	7.07	9	3.18	11	3.80
Total	361 662	831		129		57		69		90	

Cigarettes /day	Number screened	Bladder cancer deaths	Age-adjusted /10 000	All cancer deaths	Age-adjusted /10 000	CHD deaths	Age-adjusted /10 000	Stroke deaths	Age-adjusted /10 000	All cause deaths	Age-adjusted /10 000
0	228 545	16	0.67	1079	45.58	1206	50.82	112	4.73	3647	154.64
1–19	29 333	2	0.68	185	65.59	235	84.26	19	6.75	704	249.71
20–39	72 200	11	1.66	695	102.22	796	116.51	95	13.96	2309	337.19
40+	31 584	1	0.35	406	137.60	389	129.79	29	9.79	1180	394.97
Total	361 662	30		2365		2626		255		7840	

MRFIT, Multiple Risk Factor Intervention Trial; CHD, coronary heart disease

Mortality data by smoking history for the cohort

The problem of relapse creates special problems for defining smoking cessation. In most studies, cessation rates at follow-up points are cross-sectional and relate simply to the prevalence of non-smoking at that time. Such studies give no indication of the dynamics of cessation and relapse that determine the non-smoking prevalence rate, nor do they indicate what is happening in the long term with a cohort of smokers (Sherwin et al., 1981).

Ockene has analysed the cohort of smokers in MRFIT in considerable detail. This analysis demonstrated (Figure 2) that when the cohorts of SI and UC abstainers who stopped in year one are followed through the six years of the programme, 25% of the SI smokers (60% of those who stopped smoking in year one) and 6.9% (54% of those who stopped smoking in year one) of the UC smokers remained abstinent up until year six (Ockene et al., 1988).

Figure 2. Distribution of smoking cessation history at each annual visit

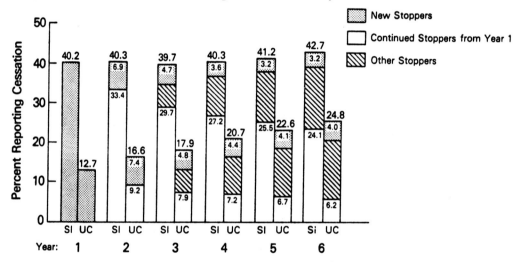

One approach that was used in the MRFIT to minimize this relapse problem in relation to end-points is to define a person who gives up smoking as an ex-smoker who reported not smoking at multiple clinic visits. For example, in Table 5, those who give up smoking are defined as previous smokers not smoking at the first and second annual examination. The mortality experience subsequent to the second annual examination was then determined.

The percentages of those who stopped smoking, using this definition, were 21.3% in the SI group (28.9% in the light smokers and 17.6% in the heavier ones) and 7.1% in the UC group (9.3% for light and 6.1% for heavier smokers).

Table 5. Mortality and cigarette smoking status at original screening and at 24 months. Mortality – number of deaths and rate per 1000 for totals

Initial smoking status[a]	Treatment group	Smoking status at 24 months		Rates of stopping smoking %	CHD		CVD		Lung cancer		All cancers		All deaths	
		Q	NQ		Q	NQ	Q	NQ	Q	NQ	Q	NQ	Q	NQ
1–29	SI	328	808	28.9	5	34	7	44	1	7	3	14	13	62
	UC	97	946	9.3	0	40	0	50	1	5	3	19	3	80
	Total	425	1754	19.5	5	74	7	94	2	12	6	33	16	142
					(11.8)	(42.2)	(16.5)	(53.6)	(4.7)	(6.8)	(14.1)	(18.8)	(37.6)	(81.0)
30+	SI	400	1877	17.6	7	50	9	78	6	26	9	51	20	163
	UC	142	2183	6.1	2	66	3	97	2	28	3	66	7	195
	Total	542	4060	11.8	9	116	12	175	8	54	12	117	27	358
					(16.6)	(28.6)	(22.1)	(43.1)	(14.8)	(13.3)	(22.1)	(28.8)	(49.8)	(88.2)
All smokers	SI	728	2685	21.3	12	84	16	122	7	33	12	65	33	225
	UC	239	3129	7.1	2	106	3	147	3	33	6	85	10	275
	Total	967	5814	14.3	14	190	19	269	10	66	18	150	43	500
					(14.5)	(32.7)	(19.6)	(46.3)	(10.3)	(11.4)	(18.6)	(25.8)	(44.5)	(86.0)

[a] Number of cigarettes smoked per day
Q, number who gave up smoking; NQ, number not giving up smoking; SI, special intervention group; UC, usual care group; CHD, coronary heart disease; CVD, cardiovascular disease
Mortality rates (in parentheses) are subsequent to the initial two years of observation

The effects of stopping smoking for two years are rather dramatic especially with regard to cardiovascular disease and total mortality. Those stopping smoking experienced a 55.7% lower CHD rate, a 57.7% lower CVD rate, a 27.9% lower total cancer mortality and a 48.3% lower total mortality compared to smokers. However, there was only a minor reduction (9.6%) from smoking cessation on lung cancer mortality even among individuals who had stopped smoking for the first two years of the trial, although the data are quite limited (Ockene et al., 1988).

Discussion

After an average of 10.5 years of follow-up, mortality rates among MRFIT men from coronary heart disease, cardiovascular disease and all causes are lower in the SI group compared to the UC group by 10.6%, 8.3% and 7.7% respectively. For the 9272 MRFIT men (72% of the original cohort) free of at-rest ECG abnormalities at the initial screening, mortality rates were lower for the SI than the UC group by 21% for coronary heart disease and 16% for all causes. For the 7841 men (61% of the cohort) with a normal ECG at rest and a normal ECG response to submaximal exercise, mortality rates were lower for the SI than the UC group by 17% for CHD and 14% for all causes. These data suggest that this multifactor intervention programme was effective in accomplishing long-term *primary* prevention of coronary, cardiovascular and all causes mortality in the SI group that had no evidence of coronary artery disease. Continued follow-up with the development of additional events should provide more definitive conclusions.

At the end of the first year of the trial, self-reported rates of stopping cigarette smoking were 43% for the SI men and 14% for the UC men; with adjustment based on serum thiocyanate levels, they were 31% and 12% respectively – an absolute difference of 19 percentage points. At year six of the trial, the self-reported rates of stopping smoking were 50% and 29%; thiocyanate-adjusted rates were 46% and 29% – an absolute difference of 17 percentage points. Moreover, data on relapse indicate that 26% of previous SI smokers and 6% of previous UC smokers had sustained stopping smoking, i.e., they had stopped smoking and did not resume the habit during the trial – a net difference of only 12 percentage points. Further, patterns of smoking cessation over time differed for the SI and the UC men, with a much higher rate of stopping smoking for SI men in the first year, then slightly higher UC than SI group rates in the next five years.

The MRFIT has been the largest attempt to measure the effects of smoking cessation on cardiovascular and total mortality in a single randomized clinical trial. The evaluation of the effect of smoking cessation were, however, confounded by the methods of selection of the participants that was based on a combination of their levels of cholesterol, blood pressure and cigarette smoking and by the fact that the interventions for all three risk factors occurred at approximately the same time. The criteria that participants be relatively healthy individuals willing to make risk factor changes and to remain in the trial for at least six years have resulted in the selection of a cohort with lower than expected mortality, especially during the first six years of initial follow-up. The MRFIT also began in 1972 at a time of increasing interest in risk factor interventions and an awareness in CHD mortality in the United States.

The results of the data from the MRFIT cohort and intervention groups are consistent in demonstrating that men who stop smoking have a substantial decrease in coronary heart disease, CVD and total mortality. The clinical trial data and SI–UC group comparisons also support these observations, but because of the multiple risk factor intervention, the results can be less confidently ascribed to any specific aspect of the intervention programme.

The limited clinical trial data with regard to cancer mortality and especially lung cancer mortality do not provide any evidence of benefit from smoking cessation. However, several observations need to be made with regard to these data: (1) the trial was not designed, especially in terms of power, for either evaluation of a single risk factor or even of multiple risk factors on cancer mortality; (2) although impressive smoking cessation rates overall were observed, a large cohort of persistent ex-smokers was not available and (3) most of the cessation occurred among the lower-risk, lighter smokers.

Consequently, the data of most value from MRFIT with regard to smoking cessation are the observational data that reaffirm the previously observed relations to disease end-points, with the notable exception of lung cancer mortality, especially among heavy smokers. The data for lung cancer in the important heavy-smoker subgroup are limited; any inferences from the data are weak, until additional information from continued follow-up can be obtained.

The MRFIT clinical trial and observational data support the benefit of multiple risk factor intervention, especially behavioural lifestyle interventions, in preventing CHD and CVD. The clinical trial data on the impact of the multiple interventions, especially smoking cessation and dietary change, on cancer mortality are too limited to allow the drawing of firm conclusions, but in general they support earlier observational data reports concerning the causal relationship between cigarette smoking and mortality from cancer of various types.

References

Benfari, R.C. (1981) Multiple Risk Factor Intervention Trial (MRFIT) III. The model for intervention. *Prev. Med.*, 10, 443–475

Caggiula, G., Christakis, M., Farrand, M., Hulley, S.B., Johnson, R., Lasser, N.L., Stamler, J. & Widdowson, G. (1981) Multiple Risk Factor Intervention Trial (MRFIT) IV. Intervention on blood lipids. *Prev. Med.*, 10, 443–475

Cohen, J.D., Grimm, R.H. & Smith, W.M. (1981) Multiple Risk Factor Intervention Trial (MRFIT) VI. Intervention on blood pressure. *Prev. Med.*, 10, 501–518

Dennis, B., Ernt, N., Hjortland, M., Tillotson, J. & Grambsch, V. (1980) The NHLBI nutrition data system. *J. Am. Diet. Assoc.*, 77, 641–647

Hughes, G.H., Hymowitz, N., Ockene, J.K., Simon, N. & Vogt, T.M. (1981) The Multiple Risk Factor Intervention Trial (MRFIT) V. Intervention on smoking. *Prev. Med.*, 10, 476–500

Multiple Risk Factor Intervention Trial Research Group (1977) Statistical design considerations in the NHLBI multiple risk factor intervention trial. *J. Chron. Dis.*, 30, 261–275

Multiple Risk Factor Intervention Trial Research Group (1982) Multiple risk factor intervention trial. Risk factor changes and mortality results. *J. Am. Med. Assoc.*, *248*, 1465–1477

Multiple Risk Factor Intervention Trial Research Group (1990) Mortality in randomized participants through 10.5 years in the multiple risk factor intervention , *J. Am. Med. Assoc.*, *263*, 1795–1801

Ockene, J.K., Kuller, L.H., Svendson, K.H. & Meilahn, E. (1988) The differential effect of smoking cessation on CHD and lung cancer mortality in the Multiple Risk Factor Intervention Trial (MRFIT): 10.5 years of follow-up (Abstract). *American Heart Association 61st Scientific Session*, November 1988

Sherwin, R., Kaelber, C.T., Kezdi, P., Kjelsberg, M.O. & Thomas, H.E., Jr (1981) The Multiple Risk Factor Intervention Trial (MRFIT) II. The development of the protocol. *Prev. Med.*, *10*, 402–425

Sherwin, R.W., Wentworth, D.N., Cutler, J.A., Hulley, S.B., Kuller, L.H. & Stamler, J. (1987) Serum cholesterol levels and cancer mortality in 361 662 men screened for the multiple risk factor intervention trial. *J. Am. Med. Assoc.*, *257*, 943–946

OTHER SELECTED INTERVENTIONS

V. Beral

*ICRF Cancer Epidemiology and Clinical Trials Unit,
Radcliffe Infirmary, Oxford, UK*

Introduction

Approaches to cancer prevention, other than those described in the preceding chapters, are varied. Many are new and the results of intervention strategies are still being evaluated. Other approaches are, by their very nature, difficult or impossible to evaluate. Some examples are discussed below.

Therapeutic agents

Drugs and diagnostic agents

Certain therapeutic and diagnostic agents such as X-rays, radionuclides, chemotherapeutic drugs, immunosuppressives and hormones have been shown to be carcinogenic. The changes in clinical practice which followed the recognition of their carcinogenicity have varied considerably. Sometimes the use of the agent had ceased, or almost ceased, by the time the carcinogenicity was demonstrated. For example, thorotrast was replaced by other contrast media even before its role in the induction of hepatic and other tumours had been noted, and the administration of hormones to pregnant women, which was common in the 1950s, was already rare in the 1970s when the increase in vaginal cancer in exposed offspring was reported. In other instances, especially where the therapeutic or diagnostic agent was highly effective and few alternatives existed, policy has generally been to minimize exposures to patients rather than to eliminate exposures altogether. The continuing use of diagnostic and therapeutic X-rays, despite their known carcinogenicity, is an example of this. Another approach has been to attempt to counter the induction and/or the progression of the malignancy by adding other agents to the therapeutic regime. For example, following the demonstration that estrogen therapy induced endometrial cancer in menopausal women, some clinicians recommended that progestogens be added to the treatment regime, since the available evidence suggested that the estrogens induced cancer because their action was 'unopposed' by progestogens.

Evaluation of the effects of such changes in clinical practice has been limited. To some extent, this reflects the complexity of the problem at hand. For example, an expected decline in cancer incidence can be demonstrated most readily when the incidence of induced cancer was very high, when the agent was used by a specific group or groups of patients, or when the agent caused cancer at an unusual site. When, however, the incidence of induced cancer was low, the exposed population was diffuse or the cancers induced were not at unusual sites, it is difficult to detect changes in cancer incidence as a result of changing clinical practice. Furthermore, the effects of partial changes in a treatment regime are more difficult to assess than major changes in therapy.

This review summarizes the few instances where the effects of changes in clinical practice have been assessed in terms of the anticipated reduction in cancer incidence.

Hormones

Some of the best examples of altered clinical practice resulting in a decline in cancer incidence relate to the use of hormonal agents.

In-utero exposure to estrogens. Hormones, usually estrogenic compounds, were widely used during the 1940s and 1950s to prevent miscarriage and to 'support' normal pregnancies. The results of clinical trials carried out in the early 1950s showed that these hormones did not reduce the risk of miscarriage nor improve the general outcome of pregnancy (Dieckmann et al., 1953; Conference on Diabetes and Pregnancy, 1955). Subsequently, the fashion for prescribing these hormones waned, the peak years of use being between 1950 and 1955 (Noller & Fish, 1974; Kinlen et al., 1974).

In 1971 the occurrence of a rare tumour, clear-cell adenocarcinoma of the vagina, was reported in teenage girls. The malignancy was found to be associated with in-utero exposure to estrogens (Herbst et al., 1971). A registry of such cases was set up, and the cumulative incidence was found to be highest in females born in 1950–1952, corresponding to the birth cohort with peak exposure to estrogens *in utero* (Melnick et al., 1987; Table 1).

Table 1. Risk of clear-cell adenocarcinoma of the vagina by year of birth (from Melnick et al., 1987)

Year of birth	Exposure index[a]	Risk per 10 000 by age 27
1947-1949	0.40	6.5
1950-1952	1.00	15.8
1953-1955	0.89	13.1
1956-1958	0.69	10.0
1959-1961	0.48	5.2

[a] Based on sales

Hormonal replacement therapy in menopausal women. The prescribing of hormones for menopausal symptoms began in the 1930s. Their use was probably relatively infrequent until the 1960s, although no reliable early statistics exist. In the United States, figures are available only for the years since 1964. These show that prescribing of premarin, the most commonly used estrogen supplement for menopausal symptoms, more than doubled between 1964 and 1975. By the mid-1970s, more than half of the postmenopausal women in California had used estrogens at some time. Estrogen use was less frequent on the east than the west coast of the United States (Jick et al., 1980; Austin & Roe, 1982; Marrett et al., 1982).

In January 1975 two reports in the *New England Journal of Medicine* described a strong association between endometrial cancer and past use of estrogens for menopausal symptoms (Smith et al., 1975; Ziel & Finck, 1975). Analyses of cancer registration data revealed that the reported incidence of endometrial cancer had, in parallel with estrogen prescriptions, doubled between 1964 and 1975. Prescribing of estrogens halved in the ensuing three years, and incidence rates of endometrial cancer in women aged 50-74 years fell correspondingly (Table 2; Jick et al., 1980; Austin & Roe, 1982; Marrett et al., 1982).

Table 2. Prescriptions of premarin in the USA and incidence of endometrial cancer in the San Francisco-Oakland metropolitan area (from Austin & Roe, 1982)

Year	Premarin prescriptions in the USA	New cases of endometrial cancer in San Francisco-Oakland metropolitan area
1968	1842	314
1969	1800	364
1970	1792	413
1971	1900	464
1972	2206	522
1973	2581	561
1974	2336	630
1975	2588	583
1976	1537	483
1977	1388	392
1978	1208	381

The rapid fall in endometrial cancer incidence since 1975 has been interpreted by some as indicative that estrogens act primarily as promoters, so that withdrawal of the drug results in a rapid decline in incidence (Austin & Roe, 1982). Marrett et al. (1982) pointed out, however, that the observed rapid increase and subsequent decrease in endometrial cancer was confined to women aged 45-64 years and to in-situ and localized malignancies. These early-stage lesions might have been selectively picked up on routine investigation of women who had bleeding problems while they were taking estrogens. Despite the decline in early-stage lesions, the incidence of non-localized endometrial cancer has continued to increase in Connecticut (Marrett et al., 1982).

In the United States, the main clinical response to the demonstrated increase in endometrial cancer associated with estrogens was to reduce the prescribing of estrogens. In the United Kingdom and elsewhere, however, a common practice has been to include progestogens in the treatment regime. The effects of the combined estrogen/progestogen regime on risk of endometrial cancer have yet to be fully evaluated.

X-rays and radionuclides

X-rays. It has been recognized since the 1950s that ionizing radiation at the doses used in clinical practice are carcinogenic. Before then, X-ray therapy was prescribed for the treatment of many benign gynaecological and cutaneous conditions such as menorrhagia, tinea capitis and enlarged thymus glands, and for chronic conditions such as ankylosing spondylitis. Over the last 40 years, increasing caution has also been exercised in the use of X-rays for diagnosis. The number of films taken and the total doses received during any specific investigation, such as a chest or an abdominal X-ray, have fallen considerably. Furthermore, X-ray treatment for benign and chronic non-malignant conditions has virtually ceased. The consequent effects, in terms of a reduction in cancer incidence, have not been investigated in adults. For childhood cancers, however, some relevant data are available.

During the 1940s and 1950s, approximately one tenth of all pregnant women underwent X-ray examination during pregnancy. Since the 1950s, the average fetal dose per film declined progressively, as well as the number of films taken per investigation (Table 3). The estimated excess childhood cancers attributed to in-utero exposures to X-rays also fell during that period, as shown in Table 3 (from Stewart & Kneale, 1970). Although the decline in estimated excess cancer risk is greater than the decline in dose, it is plausible that the more cautious use of X-rays contributed at least in part to the declining risk of childhood cancer.

Table 3. Excess risk of cancer at ages 0-9 years associated with in-utero exposure to ionizing radiation (from Stewart & Kneale, 1970)

Time period	Mean fetal dose per single film (millirads)	Estimated excess cancer deaths per million children exposed to one film
1943-1949	460	365.3
1950-1954	400	176.9
1955-1959	250	277.7
1960-1965	200	61.3

When the administration of X-rays is beneficial, such as when used in radiotherapy, the effect of withdrawing the treatment is particularly difficult to evaluate. In a review of the results of trials for breast cancer in which radiotherapy was randomized after simple or radical mastectomy, Cuzick et al. (1987) concluded that despite a similar survival in the first ten years of follow-up, thereafter women who received radiotherapy had a significant excess mortality compared to women who did not receive radiotherapy. The balance of effects may well differ for each type of cancer. Analyses such as those carried out by

Cuzick and his colleagues are perhaps the best way of determining whether the benefits of X-ray therapy outweigh its carcinogenic effects.

Radionuclides. Radionuclides are widely used in clinical practice. The doses used in diagnostic investigations are believed to be too low to be carcinogenic. An exception is thorotrast, a contrast medium used during the early 1950s, which has subsequently been shown to induce hepatic and other malignancies (da Silva Horta *et al.*, 1965). The use of thorotrast had already been replaced by alternatives before its carcinogenicity was established, and so it is difficult to demonstrate the effects of its withdrawal.

Radioactive phosphorus (^{32}P) was extensively used for the treatment of polycythaemia vera during the 1950s and early 1960s. Modan & Lilienfeld (1965) reported twenty-fold increase in leukaemia incidence in patients treated with ^{32}P, compared with patients treated by phlebotomy alone. In response to these findings, a clinical trial was set up to compare three treatments for polycythaemia vera: ^{32}P, chlorambucil and phlebotomy alone. Among the 431 patients recruited between 1967 and 1974, a significant excess of leukaemia was found in the groups treated with ^{32}P and with chlorambucil, compared with those treated by phlebotomy alone (Landaw, 1986; Table 4). Patients in the trial treated with phlebotomy alone were reported to have an excess mortality from thrombotic complication. The median overall survival in the trial participants was reported to be 11.8 years with ^{32}P treatment, 9.1 years with chlorambucil and 13.9 years with phlebotomy alone (Wasserman, 1986).

Table 4. Occurrence of acute leukaemia in patients entered into a clinical trial for treatment of polycythaemia vera (from Landaw, 1986)

Treatment	Total patients treated	Percentage of patients developing acute leukaemia up to December 1984
^{32}P	156	10.2 (16)[a]
Chlorambucil	141	13.5 (19)
Phlebotomy	134	1.5 (2)

[a] Number of cases in parentheses

Other therapeutic agents

Arsenicals and coal-tar-containing ointments cause skin cancer but the effect of replacing these medications by others has not been assessed (Stolley & Hubberd, 1982). It is well known that many drugs used in cancer therapy are themselves carcinogenic (Stolley & Hubberd, 1982). Since it is still unknown whether some drugs are more carcinogenic than others (Kaldor *et al.*, 1988), the most effective chemotherapeutic regimes remain to be assessed. Immunosuppressants are also known to be carcinogenic, but their use in transplantation is so important that therapy has not altered as a result of this observation.

Surgical procedures

There is a growing interest in the role of preventive surgery for people who are at high risk of cancer. Examples include prophylactic colectomy in patients with familial polyposis or ulcerative colitis; the removal of dysplastic naevi; and oophorectomy in women with a family history of ovarian cancer. The effectiveness of many of these procedures, especially the removal of dysplastic naevi, requires careful evaluation.

Vaccination

There is growing evidence that certain cancers are caused primarily by infectious agents. One of the best examples is hepatoma, where hepatitis B infection is known to be the main predictor of risk. Randomized trials of hepatitis B vaccination are in progress in the Gambia and China (Gambia Hepatitis Study Group, 1987). The results of these vaccine trials in terms of the prevention of hepatoma will not be available for many years.

Environmental carcinogens

Ever since the carcinogenic effects of ionizing radiation were described, legislation and voluntary controls have aimed to minimize the population's exposure to radiation – in the environment, at work and from medical investigations. The restriction of the use of ionizing radiation must have prevented cancers which could otherwise have been induced, had these restrictions not existed. But it is impossible to estimate the magnitude of the effects of such an intervention.

Legislation and voluntary controls designed to prevent or lessen the dissemination of potentially carcinogenic chemicals in the environment have, as with ionizing radiation, presumably prevented the occurrence of some cancers. Since many restrictions apply to new substances, which were never released in the environment, it is impossible to assess the effectiveness of such actions.

Only rarely has cancer occurrence been directly related to existing environmental pollutants. Even more rarely have attempts been made to control environmental pollution after such a carcinogenic effect has been demonstrated. One exception is aflatoxin contamination of food in certain African countries, although the results of control measures have not been evaluated. Reductions in levels of atmospheric ozone have been implicated as a cause of the increasing incidence of skin cancers. It has been suggested that attempts to restore ozone levels should be evaluated in terms of trends in the incidence of skin cancers. At present, however, ozone levels are continuing to fall.

Conclusions

Although it is likely that legislative and other controls of environmental contaminants have led to reductions in cancer incidence, it is often impossible to demonstrate that this is so.

The use of vaccines to prevent cancers caused by infectious agents is a new approach, well suited to rigorous evaluation. Trials of hepatitis B vaccine for hepatoma are in progress and trials for other cancers may well be instituted in the future.

The effects of changes in clinical practice designed to prevent cancers are also well suited to evaluation, although this has been rare. One example is the observed reduction in clear-cell adenocarcinomas of the vagina in the generations born after in-utero exposure to hormones had declined. The decline in the prescribing of hormones to pregnant women was, however, not designed to reduce cancer risk – it occurred before the carcinogenicity was described. The reduction in endometrial cancer incidence in the United States, following the rapid decline in prescribing of estrogens to menopausal women, is likely to be a direct consequence of specific attempts to prevent that malignancy. The substitution of one form of therapy for another does not always lead to a reduction in cancer incidence. For example, the use of the chemotherapeutic drug chlorambucil instead of radioactive phosphorus for patients with polycythaemia vera did not lead to the anticipated reduction in leukaemia risk. This illustrates the importance of evaluating the effects of changes in practice especially where one potent treatment or chemical is replaced by another. It is not known whether the experience with polycythaemia vera is an isolated episode, nor whether such examples are confined to clinical situations.

References

Austin, D.F. & Roe, K.M. (1982) The decreasing incidence of endometrial cancer: public health implications. *Am. J. Public Health*, 72, 65-68

Conference on Diabetes and Pregnancy (1955) The use of hormones in the management of pregnancy in diabetes. *Lancet*, ii, 833-836

Cuzick, J., Stewart, H., Peto, R., Baum, M., Fisher, B., Host, H., Lythgoe, J.P., Ribeiro, G., Scheurlen, H. & Wallgren, A. (1987) Overview of randomized trials of postoperative adjuvant radiotherapy in breast cancer. *Cancer Treatment Reports*, 71, 15-29

da Silva Horta, J., Abbatt, J.D., Cayolla da Motta, L.C. & Roriz, M.L. (1965) Malignancy and other late effects following administration of thorotrast. *Lancet*, ii, 201-205

Dieckmann, W.J., Davis, M.E., Rynkiewicz, L.M. & Pottinger, R.E. (1953) Does the administration of diethylstilboestrol during pregnancy have therapeutic value? *Am. J. Obstet. Gynecol.*, 66, 1062-1081

Herbst, A.L., Cole, P., Norusis, M.J., Welch, W.R. & Scully, R.E. (1979) Epidemiologic aspects and factors related to survival in 384 registry cases of clear cell adenocarcinoma of the vagina and cervix. *Am. J. Obstet. Gynecol.*, 135, 876-883

Jick, H., Walker, A.M. & Rothman, K.J. (1980) The epidemic of endometrial cancer: A commentary. *Am. J. Public Health*, 70, 264-267

Kaldor, J.M., Day, N.E. & Hemminki, K. (1988) Quantifying the carcinogenicity of antineoplastic drugs. *Eur. J. Cancer Clin. Oncol.*, 24, 703-711

Kinlen, L.J., Badaracco, M.A., Moffett, J. & Vessey, M.P. (1974) A survey of the use of oestrogens during pregnancy in the United Kingdom and of the genito-urinary cancer mortality and incidence rates in young people in England and Wales. *J. Obstet. Gynaecol. Br. Commonwealth*, 81, 849-855

Landaw, S.A. (1986) Acute leukemia in polycythemia vera. *Seminars in Hematology*, 23, 156-165

Marrett, L.D., Meigs, J.W. & Flannery, J.T. (1982) Trends in the incidence of cancer of the corpus uteri in Connecticut, 1964-1979, in relation to consumption of exogenous estrogens. *Am. J. Epidemiol.*, 116, 57-67

Melnick, S., Cole, P., Anderson, D. & Herbst, A. (1987) Rates and risks of diethylstilbestrol related clear-cell adenocarcinoma of the vagina and cervix. *New Engl. J. Med.*, 316, 514-516

Modan, B. & Lilienfeld, A.M. (1965) Polycythemia vera and leukemia - the role of radiation treatment: a study of 1222 patients. *Medicine* (Baltimore) 44, 305-344

Noller, K.L. & Fish, C.R. (1974) Diethylstilboestrol usage. *Med. Clin. North Am.*, *58*, 793-810

Smith, D.C., Prentice, R., Thompson, D.J. & Herrmann, W.L. (1975) Association of exogenous estrogen and endometrial carcinoma. *New Engl. J. Med.*, *293*, 1164-1167

Stewart, A. & Kneale, G.W. (1970) Radiation dose effects in relation to obstetric X-rays and childhood cancers. *Lancet*, *i*, 1185-1188

Stolley, D. & Hubberd, P.L. (1982) Drugs. In: Schottenfeld, P.D. & Fraumeni, J., eds, *Cancer Epidemiology and Prevention*, Philadelphia, W.B. Saunders, pp. 304-317

Wasserman, L.R. (1986) Polycythemia vera study group: a historical perspective. *Seminars in Hematology*, *23*, 183-187

Ziel, H.K. & Finkle, W.D. (1975) Increased risk of endometrial carcinoma among users of conjugated estrogens. *New Engl. J. Med.*, *293*, 1167-1170

COMMUNITY-BASED CARDIOVASCULAR DISEASE PREVENTION PROGRAMMES: MODELS FOR CANCER PREVENTION

R.A. Carleton[1,2] and T.M. Lasater[1,3]

[1]Division of Health Education,
Memorial Hospital of Rhode Island, Pawtucket, Rhode Island, USA

[2]Department of Medicine,
Memorial Hospital of Rhode Island and Brown University
Program in Medicine, Pawtucket and Providence, Rhode Island, USA

[3]Department of Community Health,
Brown University Program in Medicine,
Pawtucket and Providence, Rhode Island, USA

Atherosclerotic cardiovascular disease has been shown through numerous studies to bear close relationships with many factors (The Pooling Project Research Group, 1978). Several of these relationships have been identified through animal model research and through the methods of molecular biology (Brown et al., 1981; Brown & Goldstein, 1986), thus providing sound bases for recommending modifications of the risk factors. The impact of risk factors has been further clarified in the case of smoking, of high blood pressure, and of high cholesterol by research in which the risk factors change and in which subsequent morbid and mortal cardiovascular event rates change (Hypertension Detection and Follow-up Program Cooperative Group, 1987; Lipid Research Clinics Program, 1984; Gordon et al., 1974).

Decades of work have, in effect, established causal relationships between major risk factors and atherosclerotic disease of the coronary and other arteries. This work has laid a foundation for population-based research that aims to make changes in risk factors and, thereby, in disease rates. The pioneering efforts in this were the Stanford Three Community Study (Farquhar et al., 1977) and the North Karelia project (Hakulinen et al., this volume). Between 1977 and 1980, three major community-based primary prevention research programmes were funded by the National Heart, Lung and Blood Institute of the United States Department of Health and Human Services. All three programmes are continuing today. The Stanford Five City Project, directed by Dr John Farquhar (Farquhar et al., 1984), involves community-based education in two California cities with two others serving as control populations for risk factor surveys as well as for determining myocardial infarction and cerebrovascular accident rates. The

Minnesota Heart Health Program, under the leadership of Dr Henry Blackburn (Blackburn *et al.*, 1984), involves three education and three comparison communities in or near the state of Minnesota.

The third programme, the Pawtucket Heart Health Program (PHHP) (Lasater *et al.*, 1984; Carleton *et al.*, 1987) aims to influence the health-related behaviour of the 71 204 citizens of Pawtucket, Rhode Island. A socio-demographically comparable city is used as a reference to complete the quasi-experimental design. The independent variable in this population-based experiment, indicated to the left of Figure 1, is to involve the community through programmes designed to influence individuals, groups, organizations and the entire community. A fundamental principle, that of community involvement, helps to impart a sense of programme ownership to the target population. Media and health risk screening and counselling play important roles also.

Figure 1. The research model for the Pawtucket Heart Health Program is presented as a complex independent variable with many intervening factors influencing whether proximal or distal dependent variables are altered.

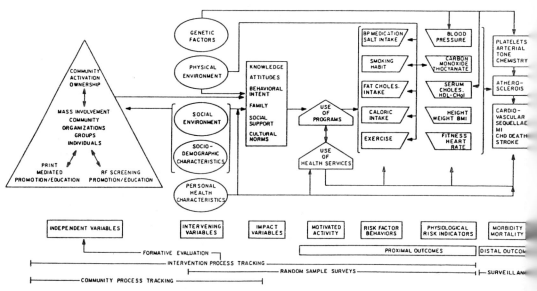

There are three categories of dependent variables. Behaviour related to risk factors, physiological risk factors themselves and incident coronary disease or cerebrovascular disease all constitute important outcomes. Some of the many factors influencing the relationships between the independent variables and the dependent variables are also indicated in Figure 1.

The process of social change occurs in stages. Correspondingly, a linked series of hypotheses is being tested by PHHP:

(*a*) a self-help approach to changes in community health, using predominantly volunteer lay facilitators is feasible;

(b) a community ownership programme will produce widespread involvement of individuals, groups, organizations and the entire community to produce health change;

(c) development of support networks will result in increased knowledge concerning risk factors, and will therefore facilitate behavioural change related to the risk factors, and help to maintain those changes;

(d) corresponding changes will occur in physiological risk factors measured in successive random samples of the Pawtucket population;

(e) the changes in physiological variables will later be manifested as a reduction in atherosclerosis-related morbidity and mortality from coronary heart disease and stroke;

(f) the changes in knowledge, in behaviour, in physiological variables, in morbidity and in mortality will occur to a greater extent in Pawtucket than in a demographically similar reference city or in other comparable reference populations.

The Pawtucket Heart Health Program targets five major modifiable cardiovascular risk factors: blood cholesterol, blood pressure, smoking prevention or cessation, physical fitness and achievement and maintenance of appropriate body weight. The process to change each of these risk factors involves individuals, groups, organizations such as work sites or churches and the entire community. Social learning theory to influence both individual and environmental factors is used as well as social marketing and behavioural psychology principles. The object is to help each citizen to change, to learn the skills necessary for change, to build a new support system, and to develop strategies for maintaining new behaviours throughout the community. Figure 2 shows the time frame of PHHP. Beginning in 1983, the programme currently has funding until the end of 1991.

Three major evaluation strategies shown in Figure 1 are also shown in Figure 2. They include random sample population surveys with both cross-sectional and cohort designs. A morbidity and mortality surveillance system and a system for following the education process are in place. The latter serves both to follow the community education programme and to obtain information back from the community.

The Pawtucket Heart Health Program targets cardiovascular disease. As this volume describes elsewhere, there is an array of evidence indicating that much of the same individual and societal behaviour is likely to influence the incidence of selected cancers. The data are most clear for the smoking habit. Considerable evidence exists, however, to suggest that altering dietary fat content may favourably affect carcinoma of the colon, breast, prostate and perhaps other sites. Although not designed to change cancer incidence, the Pawtucket Heart Health Program and other community-based primary prevention programmes offer unique opportunities to determine whether cancer incidence is influenced when lifestyles are changed.

Figure 2. The time frame of the Pawtucket Heart Health Program including major programme components and the evaluation sequence

PROGRAM YEAR	1980	1981	1982	1983	1984	1985	1986	1987	1988	1989	1990	1991
CROSS SECTION SURVEYS		XS1 2,500		XS2 2,807		XS3 2,962		XS4 2,800		XS5 2,800		
COHORT SURVEYS		XS1=C1 2,500						C2 chg. 1,540		C3 mnt. 1,240		
CHD/STROKE MORT./MORB. SURVEILLANCE		SEVEN HOSPITALS: THREE HEALTH DEPARTMENTS										
BEHAVIOR CHANGE PROGRAM							INSTITUTIONALIZATION PROCESS					
						COMMUNITY ACTIVATION						
					ORGANIZATIONAL ACTIVATION							
				SCHOOL PROGRAMS								
			PLANNING, PILOTS; PROCESS AND FORMATIVE EVALUATION									

Reproduced with permission from the *Rhode Island Medical Journal*.

To illustrate the discussion, some of the programmes designed to help people change smoking habits and nutritional patterns are briefly described. Many other aspects of the programme have been formatively tested and, if successful, implemented.

Stop-smoking programmes for small groups have been offered at worksites, in the community at large, in churches and in physician offices. Policies relating to smoking cessation have been developed and implemented at many worksites throughout the city. City-wide and worksite contests with rewards for those who stop smoking have proved successful. The cost of successive campaigns has been reduced substantially to below $50.00 per person stopping smoking.

In an effort to influence blood cholesterol levels and body weight through changes to diet, many programmes have been implemented in Pawtucket. Most grocery stores carry shelf-markers indicating lower-fat foods. Most restaurants provide a Four Heart choice on the menu. Public nutrition courses, food preparation demonstrations by local chefs and cooking programmes in Home Economics classes of schools all contribute to community education about diet and cardiovascular disease. Screening, counselling and referral events relevant to nutrition occur at least weekly within the city; at these, the food content of an individual's usual diet is assessed, blood cholesterol measured using rapid dry chemical analysers and nutrition counselling given. If necessary, referral to health professionals who are aware of the objectives of the programme is used. Weight loss classes and contests to lose weight also emphasize correct nutrition.

The PHHP has clearly documented risk factor change in many participants. By policy, changes in dependent variables will not be released until the end of the programme.

The Pawtucket Heart Health Program, and similar programmes in the United States and other countries, represent demonstration research efforts designed to learn whether population-based behaviour change can influence population disease risk and disease prevalence and incidence. Strong evidence suggests that success is feasible for cardiovascular disease. The same programmes offer hope that the burden of cancer may also be reduced. At the same time, the programmes present a potential model for the design of future primary prevention efforts for cancers that can be affected by changes to individual and collective behaviour.

Acknowledgements

The authors gratefully acknowledge the contributions of many colleagues in this work, particularly Drs Annlouise R. Assaf, R. Craig Lefebvre and Sonja M. McKinlay. This work was supported in part by Grant HL23629 from the National Heart, Lung and Blood Institute, National Institutes of Health.

References

Blackburn, H., Luepker, R., Kline, F.G., Bracht, N., Carlaw, R., Jacobs, D., Mittlemark, M., Stauffer, L. & Taylor, H.L. (1984). The Minnesota Heart Health Program: A research and demonstration project in cardiovascular disease prevention. In: Matarazzo, J.D., Weiss, S.M., Herd, A.J., Miller, N. & Weiss, S.M., eds, *Behavioral Health: A Handbook of Health Enhancement and Disease Prevention*, New York, John Wiley, pp. 1171-1178

Brown, M.S., Kovanen, P.T. & Goldstein, J.L. (1981) Regulation of plasma cholesterol by lipoprotein receptors. *Science, 212*, 628

Brown, M.S. & Goldstein, J.L. (1986) A receptor-mediated pathway for cholesterol homeostasis. *Science, 232*, 34-47

Carleton, R.A., Lasater, T.M., Assaf, A., Lefebvre, R.C. & McKinlay, S.M. (1987) The Pawtucket Heart Health Program. I. An experiment in population-based disease prevention. *Rhode Island Med. J., 70*, 533-538

Farquhar, J., Maccoby, N., Wood, P., Breitrose, H., Haskell, W., Meyer, A., Alexander, J., Brown, W., McAlister, A., Nash, J. & Stern, M. (1977) Community education for cardiovascular health. *Lancet, i*, 1192

Farquhar, J.W., Fortmann, S.P., Maccoby, N., Wood, P.D., Haskell, W.L., Taylor, C.B., Flora, V.A., Solomon, D.S., Rogers, T., Adler, E., Breitrose, P. & Weiner, L. (1984) The Stanford Five City Project: An overview. In: Matarazzo, J.D., Weiss, S.M., Herd, A.J., Miller, N. & Weiss, S.M., eds, *Behavioral Health: A Handbook of Health Enhancement and Disease Prevention*, New York, John Wiley, pp. 1154-1165

Gordon, T., Kannel, W., McGee, D. & Dawber, T. (1974) Death and coronary attacks in men after giving up cigarette smoking. A report from the Framingham Study. *Lancet, ii*, 1345-1348

Hypertension Detection and Follow-up Program Cooperative Group (1987) Five-year findings of the hypertension detection and follow-up program. I. Reduction in mortality of persons with high blood pressure including mild hypertension. *J. Am. Med. Assoc., 242*, 2562

Lasater, T., Abrams, D., Artz, L., Beaudin, P., Cabrera, L., Elder, J., Ferreira, A., Knisley, P., Peterson, G., Roderigues, A., Rosenberg, P., Snow, R. & Carleton, R.A. (1984) Lay volunteer delivery of a community-based cardiovascular risk factor change program: The Pawtucket experiment. In: Matarazzo, J.D., Weiss, S.M., Herd, A.J., Miller, N. & Weiss, S.M., eds, *Behavioral Health: A Handbook of Health Enhancement and Disease Prevention*, New York, John Wiley, pp. 1166-1170

Lipid Research Clinics Program (1984) The Lipid Research Clinics Coronary Primary Prevention Trial results. II. The relationship of reduction in incidence of coronary heart disease to cholesterol-lowering. *J. Am. Med. Assoc.*, *251*, 365–374

The Pooling Project Research Group (1978) Relationship of blood pressure, serum cholesterol, smoking habit, relative weight, and ECG abnormalities to incidence of major coronary events. Final Report of the Pooling Project. *J. Chron. Dis.*, *31*, 201–306

CURRENT ISSUES IN CANCER CHEMOPREVENTION

J.E. Buring and C.H. Hennekens

Channing Laboratory, Departments of Medicine and Preventive Medicine, Harvard Medical School and Brigham and Women's Hospital, Boston, MA, USA

The importance of identifying effective measures to prevent human cancers has been recognized for a quarter of a century (World Health Organization, 1964). The need to channel resources into prevention research is underscored by estimates that as many as 80% of all cancers in the United States could be avoided by eliminating or modifying a wide variety of exposures (Doll & Peto, 1981). Current efforts in cancer prevention are proceeding in several ways. One which is fairly well established involves investigations of environmental and lifestyle factors that could be changed in order to reduce subsequent risks of developing cancer. Another and comparatively recent approach is chemoprevention, which seeks to identify agents that can be added to an individual's usual regimen to decrease cancer risk. In this report we review current research in cancer chemoprevention and outline possible future directions.

Chemoprevention: prescription versus proscription

In the USA and other developed countries it appears that, in general, prescription of cancer inhibitors seems much more acceptable than proscription of carcinogens. Two decades of experience with anti-smoking campaigns illustrates the difficulties encountered in promoting major changes in lifestyle among populations who have not developed signs or symptoms of cancer. This situation also applies to diet, which may be implicated in approximately 35% of all malignancies (Doll & Peto, 1981). In general, modifications in diet to reduce cancer risks could involve either the elimination of foods or practices associated with increased risks (proscription), or the addition or increase to a diet of a food or nutrient that has a protective effect (prescription). Thus, in practice, the identification of possible chemopreventive agents may have an even greater public health impact than the discovery of particular cancer promoters. Moreover, micronutrients that are likely to act as inhibitors of the late stages of carcinogenesis offer particular promise as chemopreventive agents (Peto et al., 1981; Ames, 1983).

Micronutrients with chemoprevention potential

A variety of micronutrients have been identified as having the potential to decrease human cancer risk. These include dietary retinol, β-carotene or provitamin A, synthetic retinoids, vitamin B_{12}, vitamin C, vitamin E, folic acid and the trace element selenium (Ames, 1983; Hennekens et al., 1984). The available evidence concerning their possible role in cancer chemoprevention includes basic laboratory and animal research as well as observational epidemiological studies. Among these agents, however, the amount and sources of data differ markedly. For example, the chemoprevention potential of vitamins C and E derives from their theoretical ability, as antioxidants, to prevent the formation of nitrosamines and mutagens in the human digestive tract (Ames, 1983). On the basis of this property, it seemed reasonable to hypothesize that these vitamins could protect against the subsequent development of cancer. For selenium, data from descriptive epidemiological studies correlating cancer incidence rates with geographical areas of high and low concentrations of this agent (Shapiro, 1972; Schrauzer et al., 1977) have suggested its chemopreventive potential. However, data on dietary intake of this nutrient in individuals and subsequent cancer incidence are limited (Hennekens et al., 1984).

β-Carotene has broader and more solid evidence to support a role in cancer chemoprevention. Basic research has demonstrated β-carotene to be one of the most effective substances known to quench the activity of excited oxygen (Peto et al., 1981). Animal studies have demonstrated protective effects of diets rich in β-carotene against induced tumours in both mice (Epstein, 1977; Mathews-Roth, 1982; Rettura et al., 1982) and rats (Rettura et al., 1983). However, the data suggesting a chemopreventive role of β-carotene derive primarily from observational epidemiological studies in human populations. In most locations, one or two foods account for the majority of β-carotene consumed by that population, such as green and yellow vegetables in Japan, dark green leafy vegetables among the Singapore Chinese, carrots in the United States, and red palm oil in West Africa. Consequently, questionnaire studies of diet and cancer conducted in such areas have been able to examine the specific relationship of β-carotene and cancer because data on dietary intakes were collected as part of the overall investigation. Over two dozen such studies have been conducted in Europe, the USA, the Middle East and the Orient, and virtually all have showed a decreased risk of epithelial cancers among individuals having the highest intake levels compared with those having the lowest levels (Peto et al., 1981; Hennekens, 1986a, 1987). While the strongest and most consistent data concern lung cancer, protective effects have also been observed for other cancers of epithelial cell origin, including the breast and gastrointestinal tract. In addition, a number of prospective studies of serum carotene levels in relation to the development of cancer several years later have reported decreased risks among individuals who had the highest blood levels of this nutrient at baseline (Hennekens, 1986b).

The need for randomized trials

While the data suggesting a chemoprevention role for various micronutrients are generally consistent, they are hard to interpret. For example, although published studies demonstrate a clear statistical association between consumption of vegetables rich in β-carotene and decreased risks of cancer, it is not possible to distinguish whether β-carotene or some other dietary component is responsible for the observed effect. It is also possible that some other dietary or non-dietary factor associated with high intake of vegetables is reducing cancer risk among individuals eating these foods. For example, those who eat large amounts of carotene-rich foods could also be consuming less fat than those with lower intakes of β-carotene. Thus, available data have furnished promising leads but cannot provide a definitive test of the hypotheses concerning the possible roles of micronutrients in cancer chemoprevention.

Since the postulated benefits of chemopreventive agents on reducing cancer risks are small to moderate, the only way to test such hypotheses directly and reliably is by conducting randomized trials. This type of study design is uniquely suited to the evaluation of chemoprevention measures. Specifically, with a sufficiently large sample size, randomization achieves control of both known confounders, such as those mentioned above, and possible confounding variables that are either not yet known or cannot be measured. By distributing such variables equally between treatment groups, the randomized design effectively eliminates alternative explanations of any protective effect that is observed. Observational studies can do so in the design or analysis for known but not unknown confounders. As with most other chemopreventive interventions, the most plausible hypotheses concerning chemoprevention of cancer with micronutrients postulate reductions in risk of the order of 30% (Hennekens *et al.*, 1984). Whereas much larger effects, such as the 14-fold increase in risk of lung cancer or even the 80% increase in risk of coronary heart disease among cigarette smokers (Office of Smoking and Health, 1983) can be reliably demonstrated in case–control or cohort studies, effects of the order of 10–30% are about as large as the magnitude of uncontrolled confounding inherent in any observational study design.

Existing chemoprevention trials

At present, more than two dozen chemoprevention trials are in progress in the United States, China, Italy, Tanzania and Finland (Greenwald & Sondik, 1986; Greenwald *et al.*, 1987). These investigations are testing the effects of a wide variety of natural and synthetic agents on a number of cancer sites. Other than the secondary prevention trials conducted among individuals with previous skin, colon and breast cancer and oral leukoplakia, primary prevention is the objective of the following investigations:

1. Trials assessing the ability of micronutrients to halt the progression or induce regression of precancerous conditions, such as cervical dysplasia, bronchial metaplasia or dysplasia and oesophageal dysplasia.

2. Prevention trials in populations with elevated cancer risk at entry, such as cigarette smokers, asbestos workers, tin miners, albinos and individuals with familial polyposis.

3. Prevention trials in general populations of healthy individuals.

So far, only one study has reported its findings: a randomized trial designed to evaluate the effect of riboflavin, retinol and zinc on the prevalence of precancerous lesions of the oesophagus, in Huixian, People's Republic of China. After one year of follow-up, the prevalence of oesophagitis with or without dysplasia was similar in the treated and placebo groups. However, increased vitamin levels in many of the subjects in the placebo group, probably due to concurrent dietary changes, made interpretation of the results difficult. The prevalence of micronucleated oesophageal cells, which might be an indicator of very early precancerous changes was, however, lower in the treated than in the placebo group (Muñoz et al., 1985; Muñoz et al., 1987) and individuals who showed large increases in retinol, riboflavin and zinc blood levels (whether or not the change was caused by active treatment) were significantly more likely to have a histologically normal oesophagus at the end of the trial (Wahrendorf et al., 1988).

Prevention trials in the general population have the greatest potential public health importance, but such trials pose perhaps the greatest methodological obstacles in terms of recruiting and motivating large numbers of participants and following them for a sufficient length of time to accrue an adequate number of endpoints to test the hypothesis reliably. The Physicians' Health Study, an ongoing trial of β-carotene and cancer among healthy male physicians in the United States, which is also evaluating the effects of low-dose aspirin on risk of cardiovascular disease (Steering Committee et al., 1988) provides an example of the issues that must be considered in the design and conduct of a trial of cancer chemoprevention.

Timing of randomized trials

In planning a randomized trial of an agent such as β-carotene, one of the most critical issues to be addressed is timing. For both ethical and practical reasons, when such a trial is undertaken there must be sufficient doubt generally about the efficacy of the agent being tested to justify withholding it from half of the participants and, at the same time, sufficient belief in its potential to justify exposing the other half. Despite the paucity of evidence concerning the ability of one or more micronutrients to prevent cancer, their use is already widespread due in part to their availability as over-the-counter drugs. Specifically, in a cohort study of over 120 000 middle-aged nurses in the USA, approximately 35% reported regular use of multivitamin supplements in 1978 (Willett et al., 1981), and in a recent nationwide survey, 58% of US women aged 19–50 years were found to take vitamin or mineral supplements daily, compared with only 39% in 1977 (US Department of Agriculture, 1985). Sale of vitamins in the USA similarly reflects continued increases despite a lack of evidence about the health effects of their use. If supplements are not effective, their widespread use is, at best, both costly and worthless. At worst, however, a mistaken belief in their chemopreventive properties could actually be harmful if, for example, cigarette smokers took β-carotene capsules instead of stopping smoking. Consequently, the conduct of the β-carotene component of the Physicians' Health Study was timely since further delay might have made the initiation of a randomized and placebo controlled evaluation less feasible.

Use of factorial design

The Physicians' Health Study is a randomized, double-blind, placebo-controlled trial that uses a 2 × 2 factorial design to evaluate two primary prevention questions: (1) whether 325 mg of aspirin taken every other day reduces risk of cardiovascular mortality, and (2) whether 50 mg of β-carotene on alternate days reduces cancer incidence. When the trial was being planned, in the late 1970s, there was a large body of sound evidence from laboratory research, observational epidemiological studies and secondary prevention trials supporting the hypothesis that low-dose aspirin (*Lancet* editorial, 1980) reduces risks of cardiovascular disease. The β-carotene hypothesis was less mature, but, by evaluating these two research questions in a factorial design, a single trial could be used without loss of sensitivity or major increase in cost to address both questions (Stampfer *et al.*, 1985). Such an approach is an attractive option for cancer chemoprevention trials generally, since there are many promising but unproven hypotheses that could be tested simultaneously. In fact, seven of the current chemoprevention trials utilize a factorial design.

Selection of the study population

The choice of an appropriate study population is crucial to the success of a randomized trial, affecting sample size, compliance and follow-up. The decision to conduct our trial among doctors was made for several ethical, scientific and practical reasons. First of all, as a group, doctors are in an excellent position to give true, informed consent to participation in such a study. Their training permits them to recognize possible side-effects promptly and to report their medical history and health status with a greater degree of completeness and accuracy than a general population group might. Doctors are also far less mobile and easier to trace than the general population, ensuring complete follow-up information over an extended period of time. Finally, from a practical point of view, pilot studies showed that the collaboration of doctors would result in excellent compliance with their assigned regimen as well as the receipt of completed questionnaires; this allowed the trial to be conducted entirely by mail at a small fraction of the usual cost of such investigations.

In 1982, introductory letters and questionnaires were mailed to all 261 248 potentially eligible male US doctors, aged 40–84; about half returned the questionnaire, with about half of these, or 59 285, indicating that they were willing to participate in the trial. Doctors willing to participate were then excluded for the following reasons: a personal history of myocardial infarction, transient ischaemic attack or stroke, cancer, current liver or renal disease, peptic ulcer or gout; contraindications to aspirin consumption; current use of aspirin, other platelet-active drugs or non-steroidal anti-inflammatory agents; or current use of a vitamin A or β-carotene supplement. On the basis of these criteria, 33 223 of the 59 285 initially willing doctors were also eligible to participate.

The use of a run-in period to enhance compliance

One of the most important considerations in planning and conducting a valid clinical trial is the need to achieve and maintain a high level of compliance with the study regimens (Buring & Hennekens, 1984). If this does not happen, the treatment groups will resemble each other more closely, thus obscuring any true benefits of the regimen under evaluation. The effects of such dilution of treatment effects is of particular concern in chemoprevention trials attempting to distinguish reliably between small-to-moderate postulated benefits and the null hypothesis.

One strategy to enhance the likelihood of achieving high compliance is the implementation of a run-in period before randomization (Hennekens & Buring, 1987). During this period, participants are given all or part of the study regimen and only those who remain willing to continue or who demonstrate adequate adherence to the protocol after a certain amount of time are actually randomized to a study regimen. The main reason for adopting this strategy is that the largest proportion of subjects who eventually become non-compliant do so in the first months following initiation of the intervention. In the pilot studies for the Physicians' Health Study, in which participants were randomized immediately after enrolment and followed for two years, virtually all the loss in compliance occurred during the first months following enrolment. The chief reason given for not continuing was the inability to remember to take a daily pill. On the basis of this experience, we reasoned that if we enrolled 33 000 doctors and one third dropped out before being randomized, the remaining 22 000 would represent a group of proven, excellent compliers who would be much more valuable statistically to a long-term trial than 33 000 who were randomized immediately at enrolment and then failed to fulfil their commitments to the study. In this way, implementation of a run-in period would allow us to increase the power of the study substantially by yielding a group of committed participants for long-term follow-up.

Since the postulated beneficial effects of aspirin are acute and side-effects common, we wanted to expose all willing and eligible subjects to active aspirin during the run-in phase. However, the possible beneficial effects of carotene are cumulative and side-effects minimal, so that it was optimal to use a carotene placebo. Thus, the 33 223 initially willing and eligible doctors were sent calendar packs containing active aspirin and β-carotene placebo. After approximately 18 weeks, the participants were sent questionnaires; individuals who wished to discontinue participation, as well as those who developed either side-effects or an exclusion criterion, or those who wished to continue but whose compliance we deemed inadequate, were excluded from the trial before randomization. This left a total of 22 071 doctors who were then randomized into the trial.

The 22 071 doctors who remained willing and eligible after the run-in phase were randomly assigned first to two groups, one taking aspirin and the other aspirin placebo. Each of these groups was further randomized into two subgroups taking either β-carotene or its placebo. Thus, an individual participant could be receiving aspirin only, β-carotene only, both aspirin and carotene, or both placebos. All randomized doctors were sent monthly calendar packs containing a white tablet, which is either 325 mg of active aspirin or placebo, on odd-numbered days, alternating with a red capsule containing 50 mg of β-carotene or placebo, on even-numbered days. The dosage of 50 mg of β-carotene was chosen as the lowest amount that participants could take to be in the top of the distribution of β-carotene intake in the USA without experiencing yellowing of the skin. Twice a year for the first 12 months and then annually, we have sent new supplies of pills as well as brief follow-up questionnaires asking about compliance with the treatment regimens and the development of disease outcomes since the return of the last questionnaire.

Collection of pre-randomization blood samples

During the run-in phase we also asked each participating doctor to send us a blood specimen. The collection and analysis of specimens for baseline levels of retinol, carotene and retinol-binding protein was undertaken to increase the sensitivity of the trial by identifying which particular subgroup of doctors, if any, would benefit most from dietary supplementation with β-carotene. Specifically, a trial among 22 000 doctors would easily demonstrate a 30% reduction in total epithelial cancer incidence related to β-carotene if such an effect exists, but would not have sufficient power to detect a significant difference between treatment groups if the overall effect were only 10%. However, it is possible that a small but important 10% overall reduction in cancer incidence could result from a much larger effect confined exclusively to that subgroup of doctors who had low carotene or retinol levels at entry. This important public health finding could easily be picked up, given our ability to classify participants by baseline levels of these parameters, and future public health recommendations could be aimed at this type of particular subgroup. We were able to collect blood samples from 14 916 randomized doctors. These were frozen and stored in triplicate at $-85\,°C$ in three different locations; analyses are now being conducted.

Compliance and follow-up rates in the Physicians' Health Study

As the β-carotene component of this trial is still continuing, the efficacy data are unknown except to an independent Data Monitoring Board. However, there are data available concerning our goal of assembling a large cohort of compliant doctors who would be able to provide high quality, long-term follow-up data.

After an average of 4.8 years of follow-up, 87.6% of randomized subjects were still taking at least one of their two types of pills, and 83.0% were still taking both types of pills regularly. Thus, only 12.4% had stopped taking pills entirely after this extended period. These compliance figures are based solely on self-reports, since the trial is conducted entirely by mail and has participants located throughout the USA. To evaluate the reliability of this self-reported compliance, we made surprise visits to participants living in two geographical areas at four different times and collected and analysed blood and urine samples. Blood levels of β-carotene were assessed using high-pressure liquid chromatography, and there was very little overlap between the distribution of those assigned to active drug and those assigned to placebo (Satterfield, unpublished).

After 4.8 years of follow-up, 99.7% of the participating doctors had provided complete questionnaire data by either mail or telephone. This left 0.3% from whom we obtained only vital status. Thus, to date, not a single participant has been lost to mortality follow-up.

Conclusion

The ultimate goal of all these methodological issues is to refute or prove clearly the hypotheses being tested. Large-scale randomized trials that can be conducted at low cost per subject are critical to advance knowledge concerning possible cancer chemoprevention agents. However, in an area of research where there are a number of promising hypotheses, each trial must be designed and conducted in such a way that it can obtain either a definitive positive result on which public policy can be based, or a null result that is truly informative and that can safely permit the rechannelling of resources to other areas of research. Within the next few years, evidence should emerge concerning the potential of a number of micronutrients in the chemoprevention of cancer.

References

Ames, B.B. (1983) Dietary carcinogens and anticarcinogens. *Science*, 221, 1256-1264

Buring, J.E. & Hennekens, C.H. (1984) Sample size and compliance in randomized trials. In: Sestili, M.A., ed., *Chemoprevention Clinical Trials: Problems and Solutions* (NIH Publication No. 85-2715)

Doll, R. & Peto, R. (1981) *The Causes of Cancer*, Oxford, Oxford University Press

Epstein, J.H. (1977) Effects of beta-carotene on UV-induced cancer formation in the hairless mouse skin. *Photochem. Photobiol.*, 25, 211-213

Greenwald, P. & Sondik, E. (1986) Diet and chemoprevention in NCI's research strategy to achieve national cancer control objectives. *Ann. Rev. Public Health*, 7, 267-291

Greenwald, P., Cullen, J.W. & Kenna, J.W. (1987) Cancer prevention and control: From research through applications. *J. Natl Cancer Inst.*, 79, 389-400

Hennekens, C.H. (1986a) Vitamin A analogues in cancer chemoprevention. In: DeVita, V.T., Jr, Hellman, S. & Rosenberg, S.A., eds, *Important Advances in Oncology*, Philadelphia, J.B. Lippincott, pp. 23-25

Hennekens, C.H. (1986b) Micronutrients and cancer prevention. *New Engl. J. Med.*, 315, 1288-1289

Hennekens, C.H. (1987) Beta-carotene and chemoprevention of cancer. In: Cerutti, P.A., Nygaard, O.F. & Simic, M.G., eds, *Anticarcinogenesis and Radiation Protection*, New York, Plenum Press, 269-277

Hennekens, C.H. & Buring, J.E. (1987) *Epidemiology in Medicine*, Boston, Little, Brown

Hennekens, C.H., Stampfer, M.J. & Willett, W. (1984) Micronutrients and cancer chemoprevention. *Cancer Detect. Prev.*, 7, 147-158

Lancet editorial (1980) Aspirin after myocardial infarction. *Lancet*, i, 1172

Mathews-Roth, M.M. (1982) Anti-tumor activity of beta-carotene, canthaxanthin and phytoene. *Oncology*, 39, 33-37

Muñoz, N., Wahrendorf, J., Lu, J.B., Crespi, M., Day, N.E., Thurnham, D.I., Zheng, H.J., Li, B., Li, W.Y., Lin, G.L., Lan, X.Z., Correa, P., Grassi, A., O'Conor, G.T. & Bosch, F.X. (1985) No effect of riboflavin, retinol and zinc on precancerous lesions of the oesophagus. *Lancet*, ii, 111-114

Muñoz, N., Hayashi, M., Lu, J.B., Wahrendorf, J., Crespi, M. & Bosch, F.X. (1987) The effect of riboflavin, retinol and zinc on micronuclei of buccal mucosa and of oesophagus: A randomized double-blind intervention study in China. *J. Natl Cancer Inst.*, 79, 687-691

Office of Smoking and Health (1983) *Health Consequences of Smoking: Cardiovascular Disease* (Department of Health and Human Services Publication No. PHS 84-50204), Washington, DC

Peto, R., Doll, R. & Buckley, J.D. (1981) Can dietary beta-carotene materially reduce human cancer rates? *Nature*, 290, 201-208

Rettura, G., Stratford, P., Levenson, S.M. & Seifter, E. (1982) Prophylactic and therapeutic actions of supplemental β-carotene in mice inoculated with C3HBA adenocarcinoma cells. *J. Natl Cancer Inst.*, 69, 73-77

Rettura, G., Duttagupta, C., Listowsky, P., Levenson, S.M. & Seifter, E. (1983) Dimethylbenz(a)anthracene (DMBA) induced tumors: Prevention by supplemental β-carotene (abstr.). *Fed. Proc.*, *42*, 786

Satterfield, S., for the Physicians' Health Study Research Group (1989) Biochemical markers of compliance in the Physicians' Health Study. *Controlled Clinical Trials* (in press)

Schrauzer, G.N., White, D.A. & Schneider, C.J. (1977) Cancer mortality correlation studies. III. Statistical associations with dietary selenium intakes. *Bioinorg. Chem.*, *7*, 23-24

Shapiro, J.R. (1972) Selenium and carcinogenesis: a review. *Ann. NY Acad. Sci.*, *192*, 215-219

Stampfer, M., Buring, J., Willett, W., Rosner, B., Eberlein, K. & Hennekens, C.H. (1985) The 2 × 2 factorial design: its application to a randomized trial of aspirin and carotene in US physicians. *Stat. Med.*, *4*, 111-116

The Steering Committee of the Physicians' Health Study Group (Belanger, C., Buring, J.E., Eberlein, K., Goldhaber, S.Z., Gordon, D., Hennekens, C.H. (Chairman), Mayrent, S.L., Peto, R., Rosner, B., Stampfer, M., Stubblefield, F. & Willett, W.) (1988) Preliminary Report: Findings from the aspirin component of the ongoing Physicians' Health Study. *New. Engl. J. Med.*, *318*, 262-264

US Department of Agriculture (1985) *Nationwide Food Consumption Survey: Continuing Survey of Food Intakes by Individuals: Women 19-50 Years and Their Children 1-5 Years* (Nutrition Monitoring Report No. 85-1), Washington, DC

Wahrendorf, J., Muñoz, N., Lu, J.-B., Thurnham, D.I., Crespi, M. & Bosch, F.X. (1988) blood, retinol, zinc and riboflavin status in relation to precancerous lesions of the esophagus: findings from a vitamin intervention trial in the People's Republic of China. *Cancer Res.*, *48*, 2280-2283

Willett, W., Sampson, L., Bain, C., Rosner, B., Hennekens, C.H., Witschie, J. & Speizer, F.E. (1981) Vitamin supplement use among registered nurses. *Am. J. Clin. Nutr.*, *34*, 1121-1125

World Health Organization (1964) *Prevention of Cancer* (Technical Report Series 276), Geneva

SUMMARY

The Working Group[1]

As some major causes of cancer are known, the theoretical basis for the primary prevention of cancer is established. However, apart from reducing cigarette smoking, few easily applicable measures to decrease cancer incidence are available. Smoking has been estimated to cause about one third of cancer deaths in western countries (Doll & Peto, 1981; Higginson & Muir, 1979; Wynder & Gori, 1977).

Approaches to the prevention of cancer vary. Traditionally, the most common method has been health education on an individual or community basis via the mass media or by other channels. In several countries, other means have also been applied: legislative action (such as banning of advertisement), price regulation and policy decisions in relation to health services. A reduction in tobacco use has been the principal objective in most such actions. Other environmental changes have been attempted, for example, in the development of dietary policies and advice, the elimination or reduction of exposure to chemical carcinogens from the working and general environment, and the improvement of radiation protection.

Evaluation of preventive efforts is done mainly in terms of follow-up and monitoring for changes in the exposures (risk factors) believed responsible for cancer. Much less is known about the relationship between the preventive intervention and the final outcome, cancer occurrence. Efforts to evaluate the effectiveness of primary prevention measures are limited particularly because of the lack of relevant data.

Despite these difficulties, the effectiveness of primary prevention measures can be demonstrated in several circumstances. Studies of time trends in cancer in relation to time trends in exposure to risk factors, particularly tobacco smoking, are the most illuminating. International trends in lung cancer mortality, in particular, can be related to cohort-specific trends in smoking, particularly if allowance is made for the introduction of low-tar and filter cigarettes in the last 30 years. Studies of migrant populations are another well established method of demonstrating the effect which environmental change can play in the risk of cancer; however, the extent to which the change in cancer risk in migrants can be related to specific changes in their external or socio-cultural environment has not been adequately investigated.

---

[1] The members of the working group are listed on pp. ix-x

Epidemiological studies of individuals who have adopted potentially healthier habits or life-styles are relatively few and, for practical purposes, are confined to the investigation of the risk of cancer in ex-smokers. Such studies show a marked decline in risk, which is related to the time since cessation of smoking, and they constitute the most powerful evidence for the effectiveness of stopping smoking in preventing cancer.

Little is known, however, about the causes of changes in environment and life-styles at the population or individual level. Such changes may be due to educational and regulatory efforts of health authorities and to growing health-consciousness among individuals. They may also be unrelated to deliberate health-based decisions, as when food fashions change or when purely commercial factors lead to the disappearance of certain products from the market.

Sometimes intervention to remove a carcinogenic hazard from the environment has been implemented, and the effectiveness of this action can be evaluated from the study of subsequent rates of cancer incidence or mortality, despite the absence of a control group. Specific examples are found in industries where removal of some occupational exposures has resulted in the virtual elimination of known occupationally induced cancers. Clinical treatment or diagnostic practices have sometimes had carcinogenic side-effects, and when recognition of these effects led to changed practices, rates of cancer at the population level showed a subsequent decrease. For changes such as these, however, the exact design and method of intervention is rarely known.

This monograph brings together results of the relatively few studies that have been performed where specific, documented preventive interventions can be evaluated with respect to their effect on cancer risk. Most of the studies involved preventive programmes against cardiovascular disease. The interventions mainly aimed at reduction in cigarette smoking and changes in diet, the major risk factors for cardiovascular disease, and also the most important known causes of cancers. Eleven studies are described; essentially new analyses were done for seven of them and the results are described in full-length papers; four studies are covered in the more general review papers as their results relating to cancer have been previously published in some detail. Table 1 shows the basic designs of these studies, the methods of intervention and the risk-increasing or protective factors involved.

The randomized preventive trials varied in size from about 200 to 30 000 subjects. The community trials were considerable larger in size (Table 2). The period of follow-up varied from 6 to 14 years. Among the populations subjected to any type of intervention, more than 2000 cancers occurred (Table 3). The effects of the smoking and diet-related interventions on cancer risk were expected to affect mainly respiratory and gastrointestinal cancers. A summary of the findings is presented in Table 3 for all cancers, in Table 4 for lung cancer and in Table 5 for colorectal cancer.

There was little evidence of a reduced risk of total cancer in the intervention groups. The risk ratios varied between 0.5 and 1.3, with these extreme estimates derived from the smallest trials. The median of relative risk estimates was one. There was no major indication of a protective effect on the risk of lung cancer

(Table 4). The median relative risk was again close to one. None of the relative risks for colorectal cancer exceeded unity (Table 5), although relative risks less than unity were based on limited sample sizes.

All these studies have demonstrated a substantial to moderate change in intensity and prevalence of the risk factors (smoking and diet) in the target populations, indicating the success of such intervention programmes in achieving

Table 1. Main intervention studies on tobacco or diet discussed in this volume

Country, study, reference	Design	Preventive measure	Exposures considered[a]	Cancer outcome[a]
(a) Analysed in this volume				
Collaborative WHO – Italy, Poland, Spain, UK (WHO European Collaborative Group; this volume)	Group randomization by matched factories	Health education, medical services, meal services	Smoking, diet	Mortality
Finland, North Karelia (Hakulinen et al.; this volume)	Non-experimental, community trial	Health education, community and medical services	Smoking, diet	Incidence
India, Ernakulam district (Gupta et al.; this volume)	Non-experimental non-concurrent cohort study	Health education	Tobacco use	Incidence
USA MRFIT (Friedewald et al.; this volume)	Individually randomized experiment	Health education, community and medical services	Smoking, nutrition	Mortality
(b) Other studies				
Norway, Oslo (Hjermann et al., 1981)	Individually randomized experiment	Health education	Smoking diet	Mortality
Pooled study, not including Veterans Administration (Ederer et al., 1971)	Individual and group experiment	Health education, meal services	Cholesterol-lowering diet	Incidence
Sweden Gothenburg (Wilhelmsen et al., 1986)	Individually randomized experiment	Health education	Smoking cholesterol-lowering diet	Mortality
UK, Whitehall (Rose et al., 1982)	Individually randomized experiment	Health education	Smoking	Incidence & mortality
USA, Veterans Administration (Pearce & Dayton, 1971)	Individually randomized experiment	Health education, meal services	Cholesterol-lowering diet	Incidence & mortality

[a] Exposures and outcomes listed are those considered in this volume

behaviour change. It might therefore be expected that changes in risk of cancer would follow an intervention which was successful in these terms. However, the efficacy of the preventive trials in reducing the risk of cancer seems to be low. Several possibilities for the negative results should be taken into consideration, however, before such a conclusion is drawn.

Table 2. Summary of the background data of the trials

Country, study	Sex	Size of study		Length of follow-up (years)
		Intervention	Control	
Collaborative WHO –				
Italy	M	3131	2896	6
Poland	M	9110	8118	6
Spain	M	1383	1467	6
UK	M	9724	8476	6
Finland: North Karelia	M	90 500	124 000	14
	F	90 500	129 000	14
India: Oral precancer	M	7114	2858	8
	F	3069	1564	8
USA: MRFIT	M	6428	6438	10
Norway: Oslo	M	604	628	7
Pooled study	M	899	707	1–7
Sweden: Gothenburg	M	10 004	20 018	12
UK: Whitehall	M	714	731	10
USA: Veterans Administration	M	424	422	10

None of these studies was designed to examine the prevention of *cancer*. Smoking is a clear risk factor for several cancers and interventions against it are clearly appropriate. However, the relationships between diet and cancer are less firmly established. It seems probable that some of the interventions (for example, a low-fat diet) are cancer-preventing. However, a low-fat diet could imply changes in other dietary constituents which might even increase cancer risk. Some of the trials were designed to promote a diet high in polyunsaturated fat and low in saturated fat and cholesterol; relatively little is known about the effects of such dietary changes on the risk of cancer, particularly in humans. Nevertheless, it is not clear that the intervention would have been very different if cancer had been a predesigned endpoint.

Because the main focus of the studies was on the prevention of cardiovascular disease, the accuracy of the follow-up and of recording the incident cases of cancer or cancer deaths was not necessarily of the same quality as that for cardiovascular events. For the North Karelia study, however, the cancers were identified by a nationwide, population-based registration system of high reliability. Furthermore, even if less accurate, the comparability of the cancer death certificates is probably valid between the intervention and control groups, and cancer diagnoses may be even less biased (but more affected by random errors) than the diagnosis for cardiovascular deaths. It is therefore unlikely that poor quality of cancer diagnoses accounts for the limited effect.

Table 3. Summary of numbers of cancers (cases/deaths) among intervention and control groups

Country, study	Sex	Intervention	Control	Relative risk	95% Confidence interval[a]
Collaborative WHO:					
Italy	M	99	88	1.0	0.8-1.3
Poland	M	146	109	1.2	0.9-1.5
Spain	M	19	36	0.5	0.3-0.9
UK	M	300	223	1.1	0.9-1.3
Finland: North Karelia[b]	M	650	659[c]	1.0	0.9-1.1
	F	465	472[c]	1.0	0.9-1.1
India[d]: Ernakulam district	M	19[c]	20[c]	0.9	0.6-1.4
USA: MRFIT	M	140	149	0.9	0.7-1.1
Norway: Oslo	M	5	8	0.6	0.2-1.8
UK: Whitehall	M	46	36	1.3	0.8-2.0
Sweden: Gothenburg	M	315	728	0.9	0.8-1.0
Pooled study	M	11	14	0.8	0.4-1.8
USA: Veterans Administration	M	68	51	1.3	1.0-1.7

[a] Confidence intervals have been calculated from the data in the corresponding line of this table using the normal approximation to the distribution of the log relative risk
[b] Only cancers related to smoking or diet (see p. 137)
[c] Expected number
[d] Oral cancer only

Table 4. Summary of cases of lung cancer among intervention and control groups

Country, study	Sex	Intervention	Control	Relative risk	95% Confidence interval[a]
Collaborative WHO:					
Italy	M	36	24	1.4	0.8-2.3
Poland	M	42	33	1.0	0.6-1.7
Spain	M	4	7	0.6	0.2-2.0
UK	M	137	93	1.2	0.9-1.6
Finland: North Karelia	M	362	407[b]	0.9	0.8-1.1
	F	30	27[b]	1.1	0.8-1.6
USA: MRFIT	M	65	54	1.2	0.8-1.7
UK: Whitehall	M	18	24	0.8	0.4-1.5

[a] See footnote to Table 3
[b] Expected number

Table 5. Summary of cases of colorectal cancer among intervention and control groups

Country, study	Sex	Intervention	Control	Relative risk	95% Confidence interval[a]
Collaborative WHO[b]:					
Italy	M	9	10	0.8	0.3–2.0
Poland	M	10	10	1.0	0.4–2.4
Spain	M	2	4	0.3	0.1–1.6
UK	M	13	10	1.0	0.4–2.3
Finland: North Karelia	M	57	55[c]	1.0	0.8–1.3
	F	63	66[c]	1.0	0.8–1.3
USA: MRFIT	M	10	13	0.8	0.4–1.8

[a] See footnote to Table 3
[b] Including also pancreatic cancer
[c] Expected number

Cancer is a group of diseases which usually have a long latent period between first exposure and diagnosis. Where the risk factor in question acts late in the carcinogenic process, reduction in exposure might be expected to have a relatively rapid effect. For example, risk of lung cancer declines quite rapidly towards levels found in non-smokers within 15 years of smoking cessation. Migrant studies suggest that colorectal cancer risk also changes relatively rapidly after change in the environment. Risk should therefore decline in the intervention groups after the follow-up of 6 to 14 years available for the intervention trials. However, because evaluation is based on all cases of deaths within the follow-up period, the results will be greatly dominated by cases with short latent periods during the first years of follow-up. Probably only the Finnish North Karelia study had a follow-up of sufficient length to detect small effects of intervention, and in this study a total reduction in the risk of lung cancer of about 10% emerged. Furthermore, if a study is based on mortality, there is the further delay between diagnosis and death.

It should also be noted that health education in some of those studies was of relatively short duration and of low intensity. Compliance with life-style changes is inevitably far from perfect and deteriorates with time – this will be more marked with low-intensity interventions. Furthermore, most efforts were undertaken in countries where health consciousness tends to spread rapidly; it is likely that, over a longer period of time, the control populations would be exposed to similar information to that offered in the intervention group, and sometimes also to similar services. In fact, the Indian study showed a substantial and long-lasting effect of a rather limited health education campaign. Dilution of the information in the control group does not imply that such information is ineffective or unnecessary. In recent years, several community-based programmes have been started, also primarily aimed at cardiovascular disease, with continuous efforts to reduce smoking and change dietary habits by a variety of strategies. Unfortunately, data are not yet available from these studies on risk of cancer or other diseases. The potential for evaluation of such programmes, in terms of reduction of cancer incidence and other outcome indicators, should not be neglected. Although the study design may be a poor one for inferring cause and effect, existing data-collection systems, particularly cancer registries, can be used for this purpose.

In comparison with the large number of preventive projects which are implemented, relatively few investigate their effectiveness in reducing exposure, and even fewer evaluate their results in terms of the disease itself. Difficult logistics and high cost are regarded as factors inhibiting controlled preventive trials which would result in an evaluation of the effectiveness of the intervention. Nevertheless, the consequences of not carrying out randomized preventive trials should always be considered. When prevention programmes are introduced without any plan to assess their ability to detect changes in health status, the results may well be inconclusive and non-generalizable. There is a risk of repetition of such studies, which may eventually prove more costly than a few well designed trials free of bias. From the point of view of health services, there is a danger, if interventions have not been evaluated, of adopting preventive activities that are ineffective or even have adverse effects. In any case, it is likely that community intervention projects and health-related regulatory actions designed to promote healthier life-styles and environments will proliferate, without waiting for unequivocal proof of a causative link between the intervention proposed and cancer risk. It is important to design such projects in a suitable way so that an estimate can be made of their effectiveness in reducing incidence of cancer and other public health problems.

References

Doll, R. & Peto, R. (1981) *The Causes of Cancer*, Oxford, New York, Oxford University Press

Ederer, F., Leren, P., Turpeinen, O. & Frantz, I. (1971) Cancer among men on cholesterol-lowering diets. *Lancet*, ii, 203-206

Higginson, J. & Muir, C. (1979) Environmental carcinogenesis: Misconceptions and limitations to cancer control. *J. Natl Cancer Inst.*, 63, 1291-1298

Hjermann, I., Holme, I., Velve Byre, K. & Leren, P. (1981) Effect of diet and smoking intervention on the incidence of coronary heart disease: report from the Oslo Study Group of a randomized trial in healthy men. *Lancet*, ii, 1303-1310

Pearce, M. & Dayton, S. (1971) Incidence of cancer in men on a diet high in polyunsaturated fat. *Lancet*, i, 464-467

Rose, G., Hamilton, P., Colwell, L. & Shipley, M. (1982) A randomised controlled trial of anti-smoking advice: 10 year results. *J. Epidemiol. Comm. Health*, 36, 102-108

Wilhelmsen, L., Berglund, G., Elmfeld, D., Tibblin, G., Wedel, H., Pennert, K., Vedin, A., Wilhelmsson, C. & Werkö, L. (1986) The multifactor primary prevention trial in Göteborg, Sweden. *Eur. Heart J.*, 7, 279-288

Wynder, E.L. & Gori, G.B. (1977) Contribution of the environment to cancer incidence: An epidemiologic exercise. *J. Natl Cancer Inst.*, 58, 825-832

Subject index

Aflatoxin, 106-108, 176
Alcoholic drinks, 93, 102-106
American Cancer Society
    Fifty-State Study, 79, 81-83, 88
    Twenty-five State Study, 79, 81-85, 88
4-Aminobiphenyl, 24
Antimony oxide, 47
Aromatic amines, 24, 26, 44, 50, 51
Arsenic exposure, 28, 30, 39-42, 50
    in medications, 175
Asbestos
    and smoking, 51
    friction material, 38
    in gas mask manufacture, 25
    insulation material, 38, 39, 51
    textile, 37
    workers, 26, 27, 37-39
Aspirin in chemoprevention trial, 189
Australia
    alcohol consumption, 106
    immigrants, 102-105
    lung cancer, 59-65
    tobacco consumption, 72
Austria
    lung cancer, 59-62
Belgium
    lung cancer, 59-62
Benzene, 30, 31, 43
Bidi smoking, 150, 153-154
Bladder cancer
    and aromatic amines, 24, 26, 44
    and smoking, 7, 83
    in dye workers, 44
    in rubber workers, 45
    MRFIT results, 165
    North Karelia study results, 139, 140, 142, 145
Blood pressure
    reduction in CHD trials, 126, 161, 162, 181
Bone cancer, 28, 35
Breast cancer
    and vegetarian diet, 95
    fat consumption, 97, 114
    in migrant populations, 105
    meat consumption, 97
    North Karelia study results, 139-143
    screening, 5
    Women's Health Trial, 114
Breast Cancer Detection Demonstration Project, 6

Canada
    lung cancer, 60
    tobacco consumption, 72-75
Cancer
    control phases, 3-4
    registry data, 49, 96, 135
    trends, 49, 59-68, 96-99
    see also specific sites and etiological factors
Carbohydrate intake, 144
β-Carotene, 185-191
Cereal products, 93 ff., 144
Cervix uteri, cancer
    and smoking, 83
    in nuns, 24
    screening, 5
Chemoprevention, 185-192
Childhood cancers, 174
Chimney sweeps, 24, 32
Chloromethyl ethers, 28, 47
Cholesterol
    dietary reduction, 116-118, 124, 129, 159, 198
    lowered serum levels, 126, 136, 162, 181
Chromate, 30, 42-43
Cigarettes *see* Smoking, Tobacco consumption
Colon cancer
    in intervention trials, 200
    in migrant populations, 102, 104
    in Seventh-Day Adventists, 95
    MRFIT results, 163-164
    North Karelia study results, 139-142, 144, 145
    WHO Collaborative Trial results, 128-130
Compliance, 121, 190, 191
Copper smelters, 28, 30, 40-42
Cotton mill work, 25
Czechoslovakia
    lung cancer, 59-61

Dairy products and breast cancer, 97
Denmark
    lung cancer, 60-62
    tobacco consumption, 70
Diet and cancer, 7-8, 93-108, 114-121
    North Karelia study, 133
    WHO European Collaborative Trial, 123

Doctors
  chemoprevention trial, 188–191
  smoking and lung cancer, 24, 49
Drugs, 171, 175

Endometrial cancer, 175
Estrogens, 172–174

Fat
  and breast cancer, 97, 114
  intake reduction, 136, 145
  saturated vs. unsaturated, 116, 118, 159, 198
  total, 93, 116, 198
Finland
  breast cancer, 97
  lung cancer, 59–68
  North Karelia study, 133–147, 197–200
  tobacco consumption, 72, 73
France
  alcohol consumption, 105
  lung cancer, 59–65
  tobacco consumption, 72
Fruit consumption, 93 ff.
Furniture makers, 43

Gas manufacture, 25, 49
Germany, FR
  lung cancer, 59–61
Gothenburg trial, 118–120, 197–199
Greater New York Screening Program, 5

Hawaii, Japanese migrants, 101–106
Hepatitis B vaccination, 18–19, 176
Hormones, 171, 172, 177
Hungary
  lung cancer, 59–65
Hypertension see Blood pressure

Immunosuppressants, 175
India, oral precancer study, 149–156, 197–200
Intervention trials
  design, 18
  for cancer, 4, 8
  for CHD, 116, 123, 133, 157, 179, 196
  quantification of effects, 13–20
  summarized results, 196–201
Ireland
  lung cancer, 59–61
Isopropyl alcohol manufacture, 51
Italy
  lung cancer, 60, 61
  WHO Collaborative Trial results, 129

Japan
  lung cancer, 59–61, 63–65
  migrants to USA, 100–106
  prospective cohort mortality study, 79
  tobacco consumption, 72

Kidney cancer
  and smoking, 7, 83
  MRFIT results, 165

Larynx cancer
  and alcohol, 105
  and smoking, 7, 83
  rates in North Karelia study, 138–140
Latent period of cancer, 200
Leukaemia
  and benzene, 30, 31, 43
  and radium exposure, 28, 35
  and ^{32}P treatment, 175
  in radiologists, 35
Leukoplakia, 150–155
Liver cancer
  and aflatoxin, 106–108
  and thorotrast, 175
  angiosarcoma, 47–49
  HBV vaccination, 18–19, 176
Lung cancer
  and arsenic, 30, 39–42
  and asbestos, 26, 27, 37–39
  and chromate exposure, 42
  and chloromethyl ethers, 47
  and mustard gas, 25
  and smoking, 6–7, 57–90, 195, 199
    cessation of smoking, 82–88, 166
    dose-response, 68–73, 80–81
    UK doctors, 24, 49, 78
    Whitehall trial results, 114–115
  mortality data, 57 ff., 78 ff.
  MRFIT results, 163–167
  North Karelia study results, 138–40, 144
  sex differences, 79–80
  trends, 49, 59–68, 195
  WHO Collaborative Trial results, 138–140, 144

Mammography, 5
Meat consumption, 15
  and breast cancer, 97
  and colorectal cancer, 104
  and stomach cancer, 103
  in North Karelia study, 144
Mesothelioma, 37, 49
Micronutrients, 185–186
  see also Vitamins

Migrant studies, 195
  diet and cancer, 99-105
Mining
  and lung cancer, 25
  iron ore, 31
  uranium, 25
Models
  multistage carcinogenesis, 20
  statistical, in North Karelia study, 135, 146
Mozambique
  liver cancer, 107-108
MRC trial of soya-bean oil, 116, 119
MRFIT study, 120, 197
  results, 157-169, 199-200
Mule-spinning, 25
Mustard gas exposure, 25
Multiplicative effects, 16, 32, 105
Multistage carcinogenesis, 19-20
Myeloma
  in nuclear industry workers, 24

Nasal sinus cancer
  and nickel work, 31, 33-35
  and woodworking, 43
  monitoring, 49
National Cancer Act, USA, 1
Netherlands
  famine and breast cancer, 98
  lung cancer, 59-61, 63-68
  tobacco consumption, 72
Nickel exposure, 28-35, 52
North Karelia study, 133-147, 197-200
Norway
  CHD prevention trial, 118, 197
  lung cancer, 59-61
  Oslo Diet-Heart Study, 116
  tobacco consumption, 70, 72
Nuclear industry workers, 24

Obesity *see* Overweight
Occupational exposures, 23-52
Oesophageal cancer
  and alcohol, 105
  and smoking, 7, 83
  MRFIT results, 165
  North Karelia study results, 139-142
  precancerous lesions, 188
Oral cavity cancer
  and alcohol, 105
  and smoking, 7, 83
  and tobacco chewing, 149-155
  MRFIT results, 165
  precancerous lesions, 149-155

Oslo
  CHD prevention trial, 118, 197
  Diet-Heart Study, 116
Ovary cancer
  and fat and meat consumption, 97
Overweight, 124, 126, 158

Pancreatic cancer
  and smoking, 7, 83
  in Seventh-Day Adventists, 95
  MRFIT results, 165
  North Karelia study results, 139-142
  WHO Collaborative Trial results, 128-130
Pap test, 5
Pawtucket Heart Health Program, 180-183
Pesticides, arsenical, 41
Physicians' Health Study, 188-191
Poland
  WHO Collaborative Trial results, 129
Population attributable risk, 14
Portugal
  lung cancer, 59-62
Precancerous lesions, 51, 149-155, 188
Prevention studies *see* Intervention trials

Quantification of effects, 13-20

Radiation, 176
  see also following four entries and
  Nuclear industry workers, Uranium mining, X-rays
Radiologists, 35-37
Radionuclides, medical use, 171-175
Radium exposure, 28, 29, 31, 35, 36
Radon daughters, 25
Randomized prevention studies, 16, 18
Religious lifestyle, 95
  nuns, 24
  Seventh-Day Adventists, 95
Retinol, 185, 187
Risk reduction
  definitions, 14
Rubber industry, 30, 45-47

Salted foods, 93
  and stomach cancer, 101
Screening
  breast cancer, 5
  cervical cancer, 5
Scrotal cancer, 24, 25, 32, 48
Selection bias, 26, 100
Shoe workers, 31, 43
Skin cancer and radiation, 35

Smoking
   and bladder cancer, 7, 83
   and cervix cancer, 83
   and kidney cancer, 7, 83
   and larynx cancer, 7, 83
   and lung cancer, 6, 68–75, 77–90, 114–115
   and occupational risks, 32
   and oesophagus cancer, 7, 83
   and oral cavity cancer, 7, 83
   and pancreas cancer, 7, 83
   intervention trials, 114–115, 196
   reduction in CHD trials, 124, 126, 136, 158, 161–162, 181
Spain
   lung cancer, 59–62
Stomach cancer
   in Japanese migrants, 101, 102
   in migrants to Australia, 102, 103
   in Seventh-Day Adventists, 95
   North Karelia study results, 138–141, 144
   WHO Collaborative Trial results, 128–130
Surgery, preventive, 176
Swaziland
   liver cancer, 108
Sweden
   Gothenburg trial, 118–120, 197
   lung cancer, 59–61
Switzerland
   lung cancer, 59–65
Sydney Diet-Heart Study, 118

Tar content of cigarettes, 88
Thorotrast, 171, 175
Tobacco consumption, 57–90, 195
   chewed, 150–155
   trends, 68–73
   see also Smoking
Tyre factories, 45

UK
   alcohol consumption, 106
   breast cancer, 97
   doctors, 24, 49, 78, 81
   lung cancer, 59–68
   tobacco consumption, 72, 73
   Whitehall study, 113, 197
   WHO Collaborative Trial results, 128
Uranium mining, 25, 26
USA
   ACS studies, 79
   alcohol consumption, 106
   lung cancer, 60, 61, 63–68
   MRFIT study, 120, 157–169, 197–200
   tobacco consumption, 70, 72, 73
   Veteran's' study, 79

Vaccination, HBV, 18–19, 176
Vagina cancer, 171, 172
Vegetables, 93 ff., 144, 186, 187
Veterans Administration Trial (USA), 117, 197–199
Veterans' study (USA), 79, 81, 83, 84
Vinyl chloride, 47
Vitamins
   in chemoprevention, 188
   preventive effect, 145, 186

WHO European Collaborative Trial, 123–130, 197–200
Women's Health Trial, 9, 114
Woodworkers, 43, 49

X-rays
   mammography, 5
   in diagnosis and treatment, 174

PUBLICATIONS OF THE INTERNATIONAL AGENCY FOR RESEARCH ON CANCER
Scientific Publications Series

(Available from Oxford University Press through local bookshops)

No. 1 **Liver Cancer**
1971; 176 pages (*out of print*)

No. 2 **Oncogenesis and Herpesviruses**
Edited by P.M. Biggs, G. de-Thé and L.N. Payne
1972; 515 pages (*out of print*)

No. 3 **N-Nitroso Compounds: Analysis and Formation**
Edited by P. Bogovski, R. Preussman and E.A. Walker
1972; 140 pages (*out of print*)

No. 4 **Transplacental Carcinogenesis**
Edited by L. Tomatis and U. Mohr
1973; 181 pages (*out of print*)

No. 5/6 **Pathology of Tumours in Laboratory Animals, Volume 1, Tumours of the Rat**
Edited by V.S. Turusov
1973/1976; 533 pages; £50.00

No. 7 **Host Environment Interactions in the Etiology of Cancer in Man**
Edited by R. Doll and I. Vodopija
1973; 464 pages; £32.50

No. 8 **Biological Effects of Asbestos**
Edited by P. Bogovski, J.C. Gilson, V. Timbrell and J.C. Wagner
1973; 346 pages (*out of print*)

No. 9 **N-Nitroso Compounds in the Environment**
Edited by P. Bogovski and E.A. Walker
1974; 243 pages; £21.00

No. 10 **Chemical Carcinogenesis Essays**
Edited by R. Montesano and L. Tomatis
1974; 230 pages (*out of print*)

No. 11 **Oncogenesis and Herpesviruses II**
Edited by G. de-Thé, M.A. Epstein and H. zur Hausen
1975; Part I: 511 pages
Part II: 403 pages; £65.00

No. 12 **Screening Tests in Chemical Carcinogenesis**
Edited by R. Montesano, H. Bartsch and L. Tomatis
1976; 666 pages; £45.00

No. 13 **Environmental Pollution and Carcinogenic Risks**
Edited by C. Rosenfeld and W. Davis
1975; 441 pages (*out of print*)

No. 14 **Environmental N-Nitroso Compounds. Analysis and Formation**
Edited by E.A. Walker, P. Bogovski and L. Griciute
1976; 512 pages; £37.50

No. 15 **Cancer Incidence in Five Continents, Volume III**
Edited by J.A.H. Waterhouse, C. Muir, P. Correa and J. Powell
1976; 584 pages; (*out of print*)

No. 16 **Air Pollution and Cancer in Man**
Edited by U. Mohr, D. Schmähl and L. Tomatis
1977; 328 pages (*out of print*)

No. 17 **Directory of On-going Research in Cancer Epidemiology 1977**
Edited by C.S. Muir and G. Wagner
1977; 599 pages (*out of print*)

No. 18 **Environmental Carcinogens. Selected Methods of Analysis. Volume 1: Analysis of Volatile Nitrosamines in Food**
Editor-in-Chief: H. Egan
1978; 212 pages (*out of print*)

No. 19 **Environmental Aspects of N-Nitroso Compounds**
Edited by E.A. Walker, M. Castegnaro, L. Griciute and R.E. Lyle
1978; 561 pages (*out of print*)

No. 20 **Nasopharyngeal Carcinoma: Etiology and Control**
Edited by G. de-Thé and Y. Ito
1978; 606 pages (*out of print*)

No. 21 **Cancer Registration and its Techniques**
Edited by R. MacLennan, C. Muir, R. Steinitz and A. Winkler
1978; 235 pages; £35.00

No. 22 **Environmental Carcinogens. Selected Methods of Analysis. Volume 2: Methods for the Measurement of Vinyl Chloride in Poly(vinyl chloride), Air, Water and Foodstuffs**
Editor-in-Chief: H. Egan
1978; 142 pages (*out of print*)

No. 23 **Pathology of Tumours in Laboratory Animals. Volume II: Tumours of the Mouse**
Editor-in-Chief: V.S. Turusov
1979; 669 pages (*out of print*)

Prices, valid for January 1990, are subject to change without notice

List of IARC Publications

No. 24 **Oncogenesis and Herpesviruses III**
Edited by G. de-Thé, W. Henle and F. Rapp
1978; Part I: 580 pages, Part II: 512 pages (*out of print*)

No. 25 **Carcinogenic Risk. Strategies for Intervention**
Edited by W. Davis and C. Rosenfeld
1979; 280 pages (*out of print*)

No. 26 **Directory of On-going Research in Cancer Epidemiology 1978**
Edited by C.S. Muir and G. Wagner
1978; 550 pages (*out of print*)

No. 27 **Molecular and Cellular Aspects of Carcinogen Screening Tests**
Edited by R. Montesano, H. Bartsch and L. Tomatis
1980; 372 pages; £29.00

No. 28 **Directory of On-going Research in Cancer Epidemiology 1979**
Edited by C.S. Muir and G. Wagner
1979; 672 pages (*out of print*)

No. 29 **Environmental Carcinogens. Selected Methods of Analysis. Volume 3: Analysis of Polycyclic Aromatic Hydrocarbons in Environmental Samples**
Editor-in-Chief: H. Egan
1979; 240 pages (*out of print*)

No. 30 **Biological Effects of Mineral Fibres**
Editor-in-Chief: J.C. Wagner
1980; Volume 1: 494 pages; Volume 2: 513 pages; £65.00

No. 31 **N-Nitroso Compounds: Analysis, Formation and Occurrence**
Edited by E.A. Walker, L. Griciute, M. Castegnaro and M. Börzsönyi
1980; 835 pages (*out of print*)

No. 32 **Statistical Methods in Cancer Research. Volume 1. The Analysis of Case-control Studies**
By N.E. Breslow and N.E. Day
1980; 338 pages; £20.00

No. 33 **Handling Chemical Carcinogens in the Laboratory**
Edited by R. Montesano *et al.*
1979; 32 pages (*out of print*)

No. 34 **Pathology of Tumours in Laboratory Animals. Volume III. Tumours of the Hamster**
Editor-in-Chief: V.S. Turusov
1982; 461 pages; £39.00

No. 35 **Directory of On-going Research in Cancer Epidemiology 1980**
Edited by C.S. Muir and G. Wagner
1980; 660 pages (*out of print*)

No. 36 **Cancer Mortality by Occupation and Social Class 1851-1971**
Edited by W.P.D. Logan
1982; 253 pages; £22.50

No. 37 **Laboratory Decontamination and Destruction of Aflatoxins B_1, B_2, G_1, G_2 in Laboratory Wastes**
Edited by M. Castegnaro *et al.*
1980; 56 pages; £6.50

No. 38 **Directory of On-going Research in Cancer Epidemiology 1981**
Edited by C.S. Muir and G. Wagner
1981; 696 pages (*out of print*)

No. 39 **Host Factors in Human Carcinogenesis**
Edited by H. Bartsch and B. Armstrong
1982; 583 pages; £46.00

No. 40 **Environmental Carcinogens. Selected Methods of Analysis. Volume 4: Some Aromatic Amines and Azo Dyes in the General and Industrial Environment**
Edited by L. Fishbein, M. Castegnaro, I.K. O'Neill and H. Bartsch
1981; 347 pages; £29.00

No. 41 **N-Nitroso Compounds: Occurrence and Biological Effects**
Edited by H. Bartsch, I.K. O'Neill, M. Castegnaro and M. Okada
1982; 755 pages; £48.00

No. 42 **Cancer Incidence in Five Continents, Volume IV**
Edited by J. Waterhouse, C. Muir, K. Shanmugaratnam and J. Powell
1982; 811 pages (*out of print*)

No. 43 **Laboratory Decontamination and Destruction of Carcinogens in Laboratory Wastes: Some N-Nitrosamines**
Edited by M. Castegnaro *et al.*
1982; 73 pages; £7.50

No. 44 **Environmental Carcinogens. Selected Methods of Analysis. Volume 5: Some Mycotoxins**
Edited by L. Stoloff, M. Castegnaro, P. Scott, I.K. O'Neill and H. Bartsch
1983; 455 pages; £29.00

No. 45 **Environmental Carcinogens. Selected Methods of Analysis. Volume 6: N-Nitroso Compounds**
Edited by R. Preussmann, I.K. O'Neill, G. Eisenbrand, B. Spiegelhalder and H. Bartsch
1983; 508 pages; £29.00

No. 46 **Directory of On-going Research in Cancer Epidemiology 1982**
Edited by C.S. Muir and G. Wagner
1982; 722 pages (*out of print*)

List of IARC Publications

No. 47 Cancer Incidence in Singapore 1968–1977
Edited by K. Shanmugaratnam, H.P. Lee and N.E. Day
1983; 171 pages (*out of print*)

No. 48 Cancer Incidence in the USSR (2nd Revised Edition)
Edited by N.P. Napalkov, G.F. Tserkovny, V.M. Merabishvili, D.M. Parkin, M. Smans and C.S. Muir
1983; 75 pages; £12.00

No. 49 Laboratory Decontamination and Destruction of Carcinogens in Laboratory Wastes: Some Polycyclic Aromatic Hydrocarbons
Edited by M. Castegnaro, et al.
1983; 87 pages; £9.00

No. 50 Directory of On-going Research in Cancer Epidemiology 1983
Edited by C.S. Muir and G. Wagner
1983; 731 pages (*out of print*)

No. 51 Modulators of Experimental Carcinogenesis
Edited by V. Turusov and R. Montesano
1983; 307 pages; £22.50

No. 52 Second Cancers in Relation to Radiation Treatment for Cervical Cancer: Results of a Cancer Registry Collaboration
Edited by N.E. Day and J.C. Boice, Jr
1984; 207 pages; £20.00

No. 53 Nickel in the Human Environment
Editor-in-Chief: F.W. Sunderman, Jr
1984; 529 pages; £41.00

No. 54 Laboratory Decontamination and Destruction of Carcinogens in Laboratory Wastes: Some Hydrazines
Edited by M. Castegnaro, et al.
1983; 87 pages; £9.00

No. 55 Laboratory Decontamination and Destruction of Carcinogens in Laboratory Wastes: Some N-Nitrosamides
Edited by M. Castegnaro et al.
1984; 66 pages; £7.50

No. 56 Models, Mechanisms and Etiology of Tumour Promotion
Edited by M. Börzsönyi, N.E. Day, K. Lapis and H. Yamasaki
1984; 532 pages; £42.00

No. 57 N-Nitroso Compounds: Occurrence, Biological Effects and Relevance to Human Cancer
Edited by I.K. O'Neill, R.C. von Borstel, C.T. Miller, J. Long and H. Bartsch
1984; 1013 pages; £80.00

No. 58 Age-related Factors in Carcinogenesis
Edited by A. Likhachev, V. Anisimov and R. Montesano
1985; 288 pages; £20.00

No. 59 Monitoring Human Exposure to Carcinogenic and Mutagenic Agents
Edited by A. Berlin, M. Draper, K. Hemminki and H. Vainio
1984; 457 pages; £27.50

No. 60 Burkitt's Lymphoma: A Human Cancer Model
Edited by G. Lenoir, G. O'Conor and C.L.M. Olweny
1985; 484 pages; £29.00

No. 61 Laboratory Decontamination and Destruction of Carcinogens in Laboratory Wastes: Some Haloethers
Edited by M. Castegnaro et al.
1985; 55 pages; £7.50

No. 62 Directory of On-going Research in Cancer Epidemiology 1984
Edited by C.S. Muir and G. Wagner
1984; 717 pages (*out of print*)

No. 63 Virus-associated Cancers in Africa
Edited by A.O. Williams, G.T. O'Conor, G.B. de-Thé and C.A. Johnson
1984; 773 pages; £22.00

No. 64 Laboratory Decontamination and Destruction of Carcinogens in Laboratory Wastes: Some Aromatic Amines and 4-Nitrobiphenyl
Edited by M. Castegnaro et al.
1985; 84 pages; £6.95

No. 65 Interpretation of Negative Epidemiological Evidence for Carcinogenicity
Edited by N.J. Wald and R. Doll
1985; 232 pages; £20.00

No. 66 The Role of the Registry in Cancer Control
Edited by D.M. Parkin, G. Wagner and C.S. Muir
1985; 152 pages; £10.00

No. 67 Transformation Assay of Established Cell Lines: Mechanisms and Application
Edited by T. Kakunaga and H. Yamasaki
1985; 225 pages; £20.00

No. 68 Environmental Carcinogens. Selected Methods of Analysis. Volume 7. Some Volatile Halogenated Hydrocarbons
Edited by L. Fishbein and I.K. O'Neill
1985; 479 pages; £42.00

No. 69 Directory of On-going Research in Cancer Epidemiology 1985
Edited by C.S. Muir and G. Wagner
1985; 745 pages; £22.00

List of IARC Publications

No. 70 The Role of Cyclic Nucleic Acid Adducts in Carcinogenesis and Mutagenesis
Edited by B. Singer and H. Bartsch
1986; 467 pages; £40.00

No. 71 Environmental Carcinogens. Selected Methods of Analysis. Volume 8: Some Metals: As, Be, Cd, Cr, Ni, Pb, Se Zn
Edited by I.K. O'Neill, P. Schuller and L. Fishbein
1986; 485 pages; £42.00

No. 72 Atlas of Cancer in Scotland, 1975-1980. Incidence and Epidemiological Perspective
Edited by I. Kemp, P. Boyle, M. Smans and C.S. Muir
1985; 285 pages; £35.00

No. 73 Laboratory Decontamination and Destruction of Carcinogens in Laboratory Wastes: Some Antineoplastic Agents
Edited by M. Castegnaro *et al.*
1985; 163 pages; £10.00

No. 74 Tobacco: A Major International Health Hazard
Edited by D. Zaridze and R. Peto
1986; 324 pages; £20.00

No. 75 Cancer Occurrence in Developing Countries
Edited by D.M. Parkin
1986; 339 pages; £20.00

No. 76 Screening for Cancer of the Uterine Cervix
Edited by M. Hakama, A.B. Miller and N.E. Day
1986; 315 pages; £25.00

No. 77 Hexachlorobenzene: Proceedings of an International Symposium
Edited by C.R. Morris and J.R.P. Cabral
1986; 668 pages; £50.00

No. 78 Carcinogenicity of Alkylating Cytostatic Drugs
Edited by D. Schmähl and J.M. Kaldor
1986; 337 pages; £25.00

No. 79 Statistical Methods in Cancer Research. Volume III: The Design and Analysis of Long-term Animal Experiments
Edited by J.J. Gart, D. Krewski, P.N. Lee, R.E. Tarone and J. Wahrendorf
1986; 213 pages; £20.00

No. 80 Directory of On-going Research in Cancer Epidemiology 1986
Edited by C.S. Muir and G. Wagner
1986; 805 pages; £22.00

No. 81 Environmental Carcinogens: Methods of Analysis and Exposure Measurement. Volume 9: Passive Smoking
Edited by I.K. O'Neill, K.D. Brunnemann, B. Dodet and D. Hoffmann
1987; 383 pages; £35.00

No. 82 Statistical Methods in Cancer Research. Volume II: The Design and Analysis of Cohort Studies
By N.E. Breslow and N.E. Day
1987; 404 pages; £30.00

No. 83 Long-term and Short-term Assays for Carcinogens: A Critical Appraisal
Edited by R. Montesano, H. Bartsch, H. Vainio, J. Wilbourn and H. Yamasaki
1986; 575 pages; £48.00

No. 84 The Relevance of N-Nitroso Compounds to Human Cancer: Exposure and Mechanisms
Edited by H. Bartsch, I.K. O'Neill and R. Schulte-Hermann
1987; 671 pages; £50.00

No. 85 Environmental Carcinogens: Methods of Analysis and Exposure Measurement. Volume 10: Benzene and Alkylated Benzenes
Edited by L. Fishbein and I.K. O'Neill
1988; 327 pages; £35.00

No. 86 Directory of On-going Research in Cancer Epidemiology 1987
Edited by D.M. Parkin and J. Wahrendorf
1987; 676 pages; £22.00

No. 87 International Incidence of Childhood Cancer
Edited by D.M. Parkin, C.A. Stiller, C.A. Bieber, G.J. Draper. B. Terracini and J.L. Young
1988; 401 pages; £35.00

No. 88 Cancer Incidence in Five Continents Volume V
Edited by C. Muir, J. Waterhouse, T. Mack, J. Powell and S. Whelan
1987; 1004 pages; £50.00

No. 89 Method for Detecting DNA Damaging Agents in Humans: Applications in Cancer Epidemiology and Prevention
Edited by H. Bartsch, K. Hemminki and I.K. O'Neill
1988; 518 pages; £45.00

No. 90 Non-occupational Exposure to Mineral Fibres
Edited by J. Bignon, J. Peto and R. Saracci
1989; 500 pages; £45.00

No. 91 Trends in Cancer Incidence in Singapore 1968-1982
Edited by H.P. Lee, N.E. Day and K. Shanmugaratnam
1988; 160 pages; £25.00

No. 92 Cell Differentiation, Genes and Cancer
Edited by T. Kakunaga, T. Sugimura, L. Tomatis and H. Yamasaki
1988; 204 pages; £25.00

List of IARC Publications

No. 93 Directory of On-going Research in Cancer Epidemiology 1988
Edited by M. Coleman and J. Wahrendorf
1988; 662 pages (*out of print*)

No. 94 Human Papillomavirus and Cervical Cancer
Edited by N. Muñoz, F.X. Bosch and O.M. Jensen
1989; 154 pages; £19.00

No. 95 Cancer Registration: Principles and Methods
Edited by O.M. Jensen, D.M. Parkin, R. MacLennan, C.S. Muir and R. Skeet
Publ. due 1990; approx. 300 pages

No. 96 Perinatal and Multigeneration Carcinogenesis
Edited by N.P. Napalkov, J.M. Rice, L. Tomatis and H. Yamasaki
1989; 436 pages; £48.00

No. 97 Occupational Exposure to Silica and Cancer Risk
Edited by L. Simonato, A.C. Fletcher, R. Saracci and T. Thomas
1990; 124 pages; £19.00

No. 98 Cancer Incidence in Jewish Migrants to Israel, 1961-1981
Edited by R. Steinitz, D.M. Parkin, J.L. Young, C.A. Bieber and L. Katz
1989; 320 pages; £30.00

No. 99 Pathology of Tumours in Laboratory Animals, Second Edition, Volume 1, Tumours of the Rat
Edited by V.S. Turusov and U. Mohr
Publ. due 1990; 740 pages; £85.00

No. 100 Cancer: Causes, Occurrence and Control
Editor-in-Chief L. Tomatis
1990; 352 pages; £24.00

No. 101 Directory of On-going Research in Cancer Epidemiology 1989-90
Edited by M. Coleman and J. Wahrendorf
1989; 818 pages; £36.00

No. 102 Patterns of Cancer in Five Continents
Edited by S.L. Whelan and D.M. Parkin
1990; 162 pages; £25.00

No. 103 Evaluating Effectiveness of Primary Prevention of Cancer
Edited by M. Hakama, V. Beral, J.W. Cullen and D.M. Parkin
Publ. due 1990; approx. 250 pages; £32.00

No. 104 Complex Mixtures and Cancer Risk
Edited by H. Vainio, M. Sorsa and A.J. McMichael
Publ. due 1990; 442 pages; £38.00

No. 105 Relevance to Human Cancer of *N*-Nitroso Compounds, Tobacco Smoke and Mycotoxins
Edited by I.K. O'Neill, J. Chen, S.H. Lu and H. Bartsch
Publ. due 1990; approx. 600 pages

List of IARC Publications

IARC MONOGRAPHS ON THE EVALUATION OF CARCINOGENIC RISKS TO HUMANS
(Available from booksellers through the network of WHO Sales Agents*)

Volume 1 Some Inorganic Substances, Chlorinated Hydrocarbons, Aromatic Amines, *N*-Nitroso Compounds, and Natural Products
1972; 184 pages (*out of print*)

Volume 2 Some Inorganic and Organometallic Compounds
1973; 181 pages (out of print)

Volume 3 Certain Polycyclic Aromatic Hydrocarbons and Heterocyclic Compounds
1973; 271 pages (*out of print*)

Volume 4 Some Aromatic Amines, Hydrazine and Related Substances, *N*-Nitroso Compounds and Miscellaneous Alkylating Agents
1974; 286 pages;
Sw. fr. 18.-/US $14.40

Volume 5 Some Organochlorine Pesticides
1974; 241 pages (*out of print*)

Volume 6 Sex Hormones
1974; 243 pages (*out of print*)

Volume 7 Some Anti-Thyroid and Related Substances, Nitrofurans and Industrial Chemicals
1974; 326 pages (*out of print*)

Volume 8 Some Aromatic Azo Compounds
1975; 375 pages;
Sw. fr. 36.-/US $28.80

Volume 9 Some Aziridines, *N*-, *S*- and *O*-Mustards and Selenium
1975; 268 pages;
Sw.fr. 27.-/US $21.60

Volume 10 Some Naturally Occurring Substances
1976; 353 pages (*out of print*)

Volume 11 Cadmium, Nickel, Some Epoxides, Miscellaneous Industrial Chemicals and General Considerations on Volatile Anaesthetics
1976; 306 pages (*out of print*)

Volume 12 Some Carbamates, Thiocarbamates and Carbazides
1976; 282 pages;
Sw fr. 34.-/US $27.20

Volume 13 Some Miscellaneous Pharmaceutical Substances
1977; 255 pages;
Sw. fr. 30.-/US$ 24.00

Volume 14 Asbestos
1977; 106 pages (*out of print*)

Volume 15 Some Fumigants, The Herbicides 2,4-D and 2,4,5-T, Chlorinated Dibenzodioxins and Miscellaneous Industrial Chemicals
1977; 354 pages;
Sw. fr. 50.-/US $40.00

Volume 16 Some Aromatic Amines and Related Nitro Compounds - Hair Dyes, Colouring Agents and Miscellaneous Industrial Chemicals
1978; 400 pages;
Sw. fr. 50.-/US $40.00

Volume 17 Some *N*-Nitroso Compounds
1987; 365 pages;
Sw. fr. 50.-/US $40.00

Volume 18 Polychlorinated Biphenyls and Polybrominated Biphenyls
1978; 140 pages;
Sw. fr. 20.-/US $16.00

Volume 19 Some Monomers, Plastics and Synthetic Elastomers, and Acrolein
1979; 513 pages;
Sw. fr. 60.-/US $48.00

Volume 20 Some Halogenated Hydrocarbons
1979; 609 pages (*out of print*)

Volume 21 Sex Hormones (II)
1979; 583 pages;
Sw. fr. 60.-/US $48.00

Volume 22 Some Non-Nutritive Sweetening Agents
1980; 208 pages;
Sw. fr. 25.-/US $20.00

Volume 23 Some Metals and Metallic Compounds
1980; 438 pages (*out of print*)

Volume 24 Some Pharmaceutical Drugs
1980; 337 pages;
Sw. fr. 40.-/US $32.00

Volume 25 Wood, Leather and Some Associated Industries
1981; 412 pages;
Sw. fr. 60.-/US $48.00

Volume 26 Some Antineoplastic and Immunosuppressive Agents
1981; 411 pages;
Sw. fr. 62.-/US $49.60

Volume 27 Some Aromatic Amines, Anthraquinones and Nitroso Compounds, and Inorganic Fluorides Used in Drinking Water and Dental Preparations
1982; 341 pages;
Sw. fr. 40.-/US $32.00

Volume 28 The Rubber Industry
1982; 486 pages;
Sw. fr. 70.-/US $56.00

Volume 29 Some Industrial Chemicals and Dyestuffs
1982; 416 pages;
Sw. fr. 60.-/US $48.00

List of IARC Publications

Volume 30 Miscellaneous Pesticides
1983; 424 pages;
Sw. fr. 60.-/US $48.00

Volume 31 Some Food Additives, Feed Additives and Naturally Occurring Substances
1983; 314 pages;
Sw. fr. 60.-/US $48.00

Volume 32 Polynuclear Aromatic Compounds, Part 1: Chemical, Environmental and Experimental Data
1984; 477 pages;
Sw. fr. 60.-/US $48.00

Volume 33 Polynuclear Aromatic Compounds, Part 2: Carbon Blacks, Mineral Oils and Some Nitroarenes
1984; 245 pages;
Sw. fr. 50.-/US $40.00

Volume 34 Polynuclear Aromatic Compounds, Part 3: Industrial Exposures in Aluminium Production, Coal Gasification, Coke Production, and Iron and Steel Founding
1984; 219 pages;
Sw. fr. 48.-/US $38.40

Volume 35 Polynuclear Aromatic Compounds: Part 4: Bitumens, Coal-tars and Derived Products, Shale-oils and Soots
1985; 271 pages;
Sw. fr. 70.-/US $56.00

Volume 37 Tobacco Habits Other than Smoking: Betel-quid and Areca-nut Chewing; and some Related Nitrosamines
1985; 291 pages;
Sw. fr. 70.-/US $56.00

Volume 38 Tobacco Smoking
1986; 421 pages;
Sw. fr. 75.-/US $60.00

Volume 39 Some Chemicals Used in Plastics and Elastomers
1986; 403 pages;
Sw. fr. 60.-/US $48.00

Volume 40 Some Naturally Occurring and Synthetic Food Components, Furocoumarins and Ultraviolet Radiation
1986; 444 pages;
Sw. fr. 65.-/US $52.00

Volume 41 Some Halogenated Hydrocarbons and Pesticide Exposures
1986; 434 pages;
Sw. fr. 65.-/US $52.00

Volume 42 Silica and Some Silicates
1987; 289 pages;
Sw. fr. 65.-/US $52.00

Volume 43 Man-Made Mineral Fibres and Radon
1988; 300 pages;
Sw. fr. 65.-/US $52.00

Volume 44 Alcohol Drinking
1988; 416 pages;
Sw. fr. 65.-/US $52.00

Volume 45 Occupational Exposures in Petroleum Refining; Crude Oil and Major Petroleum Fuels
1989; 322 pages;
Sw. fr. 65.-/US $52.00

Volume 46 Diesel and Gasoline Engine Exhausts and Some Nitroarenes
1989; 458 pages;
Sw. fr. 65.-/US $52.00

Volume 47 Some Organic Solvents, Resin Monomers and Related Compounds, Pigments and Occupational Exposures in Paint Manufacture and Painting
1990; 536 pages;
Sw. fr. 85.-/US $68.00

Volume 48 Some Flame Retardants and Textile Chemicals, and Exposures in the Textile Manufacturing Industry
1990; 345 pages;
Sw. fr. 65.-/US $52.00

Volume 49 Chromium, Nickel and Welding
1990; 677 pages;
Sw. fr. 95.–/US$76.00

Supplement No. 1 Chemicals and Industrial Processes Associated with Cancer in Humans (IARC Monographs, Volumes 1 to 20)
1979; 71 pages; (*out of print*)

Supplement No. 2 Long-term and Short-term Screening Assays for Carcinogens: A Critical Appraisal
1980; 426 pages;
Sw. fr. 40.-/US $32.00

Supplement No. 3 Cross Index of Synonyms and Trade Names in Volumes 1 to 26
1982; 199 pages (*out of print*)

Supplement No. 4 Chemicals, Industrial Processes and Industries Associated with Cancer in Humans (IARC Monographs, Volumes 1 to 29)
1982; 292 pages (*out of print*)

Supplement No. 5 Cross Index of Synonyms and Trade Names in Volumes 1 to 36
1985; 259 pages;
Sw. fr. 46.-/US $36.80

Supplement No. 6 Genetic and Related Effects: An Updating of Selected IARC Monographs from Volumes 1 to 42
1987; 729 pages;
Sw. fr. 80.-/US $64.00

Supplement No. 7 Overall Evaluations of Carcinogenicity: An Updating of IARC Monographs Volumes 1–42
1987; 434 pages;
Sw. fr. 65.-/US $52.00

Supplement No. 8 Cross Index of Synonyms and Trade Names in Volumes 1 to 46 of the IARC Monographs
Publ. due 1990; 260 pages;
Sw. fr. 60.-/US $48.00

List of IARC Publications

IARC TECHNICAL REPORTS*

No. 1 **Cancer in Costa Rica**
Edited by R. Sierra,
R. Barrantes, G. Muñoz Leiva,
D.M. Parkin, C.A. Bieber and
N. Muñoz Calero
1988; 124 pages;
Sw. fr. 30.-/US $24.00

No. 2 **SEARCH: A Computer Package to Assist the Statistical Analysis of Case-control Studies**
Edited by G.J. Macfarlane,
P. Boyle and P. Maisonneuve (in press)

No. 3 **Cancer Registration in the European Economic Community**
Edited by M.P. Coleman and
E. Démaret
1988; 188 pages;
Sw. fr. 30.-/US $24.00

No. 4 **Diet, Hormones and Cancer: Methodological Issues for Prospective Studies**
Edited by E. Riboli and
R. Saracci
1988; 156 pages;
Sw. fr. 30.-/US $24.00

No. 5 **Cancer in the Philippines**
Edited by A.V. Laudico,
D. Esteban and D.M. Parkin
1989; 186 pages;
Sw. fr. 30.-/US $24.00

No. 6 **La genèse du Centre International de Recherche sur le Cancer**
Par R. Sohier et A.G.B. Sutherland
1990; 104 pages
Sw. fr. 30.-/US $24.00

No. 7 **Epidémiologie du cancer dans les pays de langue latine**
1990; 310 pages
Sw. fr. 30.-/US $24.00

No. 8 **Comparative Study of Anti-smoking Legislation in Countries of the European Economic Community**
Edited by A. Sasco
1990; c. 80 pages
Sw. fr. 30.-/US $24.00
(English and French editions available)

DIRECTORY OF CHEMICALS BEING TESTED FOR CARCINOGENICITY
(Until Vol. 13 Information Bulletin on the Survey of Chemicals Being Tested for Carcinogenicity)*

No. 8 Edited by M.-J. Ghess,
H. Bartsch and L. Tomatis
1979; 604 pages; Sw. fr. 40.-

No. 9 Edited by M.-J. Ghess,
J.D. Wilbourn, H. Bartsch and
L. Tomatis
1981; 294 pages; Sw. fr. 41.-

No. 10 Edited by M.-J. Ghess,
J.D. Wilbourn and H. Bartsch
1982; 362 pages; Sw. fr. 42.-

No. 11 Edited by M.-J. Ghess,
J.D. Wilbourn, H. Vainio and
H. Bartsch
1984; 362 pages; Sw. fr. 50.-

No. 12 Edited by M.-J. Ghess,
J.D. Wilbourn, A. Tossavainen
and H. Vainio
1986; 385 pages; Sw. fr. 50.-

No. 13 Edited by M.-J. Ghess,
J.D. Wilbourn and A. Aitio
1988; 404 pages; Sw. fr. 43.-

No. 14 Edited by M.-J. Ghess,
J.D. Wilbourn and H. Vainio
1990; c. 400 pages;
Sw. fr. 45.-

NON-SERIAL PUBLICATIONS †

Alcool et Cancer
By A. Tuyns (in French only)
1978; 42 pages; Fr. fr. 35.-

Cancer Morbidity and Causes of Death Among Danish Brewery Workers
By O.M. Jensen 1980;
143 pages; Fr. fr. 75.-

Directory of Computer Systems Used in Cancer Registries
By H.R. Menck and D.M. Parkin 1986; 236 pages;
Fr. fr. 50.-

---

* Available from booksellers through the network of WHO sales agents.

† Available directly from IARC